Welding Fabrication

VOLUME I

Welding Fabrication

VOLUME I

TECHNOLOGICAL BACKGROUND

J. G. Tweeddale

F.I.M., M.I.Mech.E., M.Inst.W.

Senior Lecturer in Fabrication Metallurgy
Imperial College, London

LONDON ILIFFE BOOKS LTD
AMERICAN ELSEVIER PUBLISHING COMPANY, INC.
NEW YORK   1969

First published in 1969 by
Iliffe Books Limited
42 Russell Square, London, W.C.1

© J. G. Tweeddale, 1969

Published in the U.S.A. by
American Elsevier Publishing Company, Inc.
52 Vanderbilt Avenue, New York, N.Y. 10017

444-19756-7

Printed in England

Contents

Preface

The technician who wishes to learn something of the technological background of welding finds himself in rather a quandary. There are publications which purport to cover the subject but these tend to group themselves under one of two headings, namely, craft practice or specialised application, without giving the broad simple treatment so necessary for a genuine understanding.

This present work is an attempt to produce a simple reliable text book to fill this gap at a sufficiently high level of treatment but at relatively low cost. The treatment is qualitative rather than quantitative and is condensed to the minimum. No attempt is made to go deeply into particular theories or practices. On the other hand, no essential basic principle is omitted thus, for particular branches of technology such as metallurgy, the principal aspects are summarised so that the relevant factors may be seen in relation to the main theme.

Since the main aim is to help apprentice technicians, technician students and apprentice craftsmen, the subject matter is subdivided into three parts:

1. Technological Background
2. Processes
3. Fabrication

Each part is published separately to keep physical size down and to make purchase in easy stages a possibility; but it should not be forgotten that the three parts are closely interrelated and cross-referenced with each other.

It is believed that the coverage of the subject is adequate for the relevant C.G.L.I., H.N.D. and undergraduate courses requiring knowledge of welding technology as a subsidiary part of the main study and also for the earlier stages of courses for welding engineers.

Thanks are due to the many friends who have helped and advised.

Imperial College, London, 1969   J.G.T.

CHAPTER 1

Introduction

1.1 DEFINITION OF WELDING

Many attempts have been made to define the term 'welding' but it is
not easy to be precise since its scope includes so many fields. Basically,
a weld is a metallurgical operation but it may be performed on many
materials in a variety of very different ways and in widely differing
situations, for example, in the joining of the tiny components of a
miniature radio valve, the fabrication of motor car bodies, manufacture
of aircraft components, site welding of buildings and bridges and construction of giant reactor pressure vessels for atomic energy applications. There are also several ancillary processes such as flame cutting
and metal spraying that are almost inseparable from the main field of
welding.

In British Standard 499: Part 1: 1965, Welding Terms and Symbols,
a weld is defined as 'A union between pieces of metal at faces rendered
plastic or liquid by heat, or by pressure, or both', with the note added:
'Filler metal may be used to effect the union'. This is probably the
most reasonable definition that can be expected in the circumstances.
However, welding principles are also applied, in a similar way, to the
fabrication of many synthetic thermoplastic materials.

Much of the obscurity about welding probably arises from a glib
acceptance of welding as just another tool of fabrication, without
knowing or understanding, the great variety of technologies that may
have been used to make that tool effective. In the following pages,
the technology of welding will be surveyed in sufficient detail to give a

1

broad picture of the possibilities and limitations of the various processes.

1.2   HISTORICAL BACKGROUND TO WELDING

Although the metallurgical nature of most welding processes is now well established, the basic methods used for achieving a weld were developed on an empirical basis over many centuries of craft practice.

The earliest definitely known example of a weld is a brazed joint in a piece of copper panelling, dating from about 3000 B.C., found in Mesopotamia. Examples of solid-phase welds of the blacksmith type are known, from about 1350 B.C. in the Middle East (one Egyptian example having been found in King Tutankhamen's tomb), from about A.D. 310 in India (e.g., the Delhi Pillar) and from about A.D. 400 in Europe. Full fusion welding was not practised regularly until the 19th century, owing to the lack of a concentrated heat source, although some very early examples (500 B.C.) of possible 'burning-on' joining of cast parts and repairing of defective castings do exist.

Prior to the Industrial Revolution, it was the need for armaments that gave most incentive to the development of what was then a mysterious branch of the blacksmith's art and many beautiful examples of welded chain link armour, welded swords and welded metal inlaying of armour still survive in various parts of the world. However, it was the Industrial Revolution that put the first spur to development of welding in civil applications; thus, the invention of the oxy-acetylene blowpipe was quickly adapted to fabrication and repair. Adaptation of electric arc and resistance heating was not so speedy until a fresh military spur was applied by the Great War of 1914–18 and later by the World War of 1939–45. In fact, welding practice and theory has advanced more since 1939 than in the whole of its previous history. Now, what was previously a jealously secretive art has developed into an established, progressive technology that embraces every field of engineering.

1.3   THE METALLURGICAL BASIS TO WELDING

In the course of the rapid development mentioned above, it is not surprising that basic technology tended to be forgotten and processes were regarded from the mechanistic aspect. Hence, it became the custom to class welding processes loosely under two main headings, namely, 'fusion' and 'pressure' (see Vol. 2, Chapter 1); but this classification is much too vague to be of any use. The overwhelming majority of welding applications are based on metallurgical principles, so it is most sensible to look at them in that connotation. Thus, welding processes can fall into one of the following classes, according to the diffusion bonding mode (see Vol. 2, Chapter 1):

1. Liquid to liquid diffusion (full fusion welds).
2. Liquid to solid diffusion (brazing).
3. Solid to solid diffusion (solid phase welding).

To understand this grouping, knowledge of the crystalline nature of metals is required.

1.3.1 *Metallurgy and Welding Technology*

Many disciplines can and do contribute to an understanding of welding technology but none contribute as much as metallurgy and a detailed knowledge of it is essential for good understanding of the principles of effective welding. For complete knowledge of the welding possibilities of a particular metallic material, a metallurgical training is a necessity; but the engineer with occasional interest in welding, or most of his interest in the manufacturing applications, can operate effectively without detailed understanding, provided that he has (*a*) sufficient metallurgical knowledge to enable him to judge when specialist knowledge or advice is required and (*b*) sufficient intelligence to seek that specialist help when it is needed.

BIBLIOGRAPHY

TYLECOTE, R. F., *Metallurgy in Archeology*, Edward Arnold (1962).
WINTERTON, K., *Welding Metal Fabric.*, 438, Nov. (1962); 488, Dec. (1962); 71 Feb. (1963).

CHAPTER 2
Basic Metallurgy

2.1 THE CRYSTALLINE NATURE OF A METAL

The constituent atoms of a solid metal tend to position themselves in accordance with a simple but strict and regular geometric 'space lattice' pattern, which repeats itself throughout the solid mass so that each atom or group of atoms is surrounded by exactly the same array of neighbouring atoms as every other atom or group. This systematic array constitutes a crystal structure and the basic pattern a unit 'cell' of the structure.

Fourteen types of cell pattern are geometrically possible but three types predominate in the common structural metallic materials, and knowledge of these three is all that is required for most purposes. Although several materials may have the same type of space lattice, the lattice in each material will be characteristic in unit size and properties of the atoms of which it is composed.   Several metals are 'allotropic', that is they are capable of existing with different lattice configuration under differing conditions and may change their atomic pattern at a particular temperature level.   A few common metals have this property, notably iron (Fe), and it may make them particularly suitable for certain types of metallurgical treatment, see Section 2.6.

The three dominant types of unit cell are:

1. Close-packed hexagonal (or hexagonal, close-packed).
2. Face-centred cubic.
3. Body-centred cubic.

Each is the outcome of the constituent atoms trying to arrange themselves together, as closely as the nature of their physical and chemical properties will allow, to form a 'metallic bond' within the crystal. A metallic bond is one in which certain of the electrons of each atom are mutually shared throughout the crystal mass; a group of positively charged atoms being held in position within a negatively charged electron 'cloud'. For further details of these principles, reference should be made to a book such as *Metallurgical Principles for Engineers* by the present author (Iliffe Books Ltd.); or to a textbook of metallurgy or of physical chemistry. It is this metallic bond that gives to metals their relatively high thermal and electrical conductivities.

1. A *close-packed hexagonal (CPH or HCP) lattice* is made up of a single system of 'close-packed' planes of atoms in which each atom is as close to as many other atoms as it can be. That is, the atoms in the plane are positioned at the apices of a linked system of equal

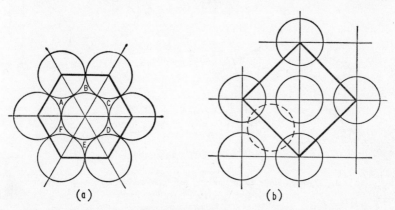

(a)   (b)

2.1. *Modes of packing of atoms in planes of crystals. (a) 'Close hexagonal' packing. (b) 'Square' packing*

sized equilateral triangles and each atom is surrounded by six equally distant close neighbours in the same plane, see Fig. 2.1 (a). The successive planes of atoms lie one above the other nestling into each other in the appropriate one of two possible alternative positions so that every *alternate* plane is in vertical alignment as shown in Fig. 2.2 (a) and each atom has three

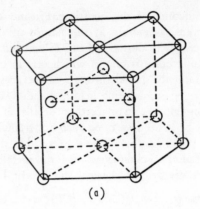

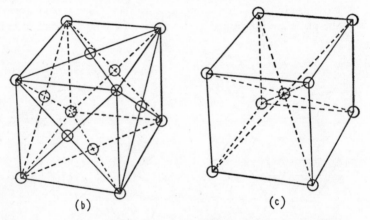

Fig. 2.2. Common space lattice arrays. (a) Close-packed hexagonal. (b) Face-centred cubic. (c) Body-centred cubic

equally distant close neighbours in each adjacent plane, making twelve equally distant close neighbours in all.

2. A *face-centred cubic (FCC) lattice* is similarly made up; but only every *third* plane is in vertical alignment by occupying the appropriate alternative nestling positions for successive planes one above the other. As a result, the atoms are positioned as shown in Fig. 2.2 (b), the perpendicular axis to a set of close-packed planes running through diametrically opposite corners of the cube.

There are four discernable systems of close packed planes forming so-called octahedral planes. There are twelve equally spaced close neighbours to each atom as in the CPH system.

3. A *body-centred cubic (BCC) lattice* differs in that the atoms in the face planes of the unit cube array themselves on a 'square' pattern as shown in Fig. 2.1 (b), each with four equally spaced neighbours. Successive planes nestle in the one available nestling position to give each atom eight equally spaced closest neighbours, four from each *adjacent* plane. It should be noted that the atoms closest to each other within any one face plane are not quite so close as neighbouring atoms in adjacent planes.

These configurations of atoms can have marked effects on the mechanical properties of the metals in which they occur. It should be noted that each of these systems can be seen from certain different directions to be made up of stacks of planes of atoms with other systems of regular spacing within the planes, on the lines suggested in the description of the FCC lattice. Such planes of regularly spaced atoms constitute what are known as 'crystallographic planes'.

2.1.1 *Polycrystalline Structure of a Metal*

Ideally, any mass of metal could be imagined as consisting of one large crystal or 'molecule', and it is possible to 'grow' very large single crystals; but generally, the nature of the solidification process, by which a crystalline mass is usually formed, prevents this happening and the metal is composed of a 'polycrystalline aggregate' or 'polycrystal' in which there is a large number of differently orientated distinguishable crystals or 'grains', closely bonded to each other.

Nearly all engineering metals and alloys are formed initially by solidification from the liquid state. In the liquid state, the atoms are constantly moving by sliding past each other, since the 'liquid bond' will hold atoms together but the high thermal energy keeps them intermingling freely and moving past each other (or 'diffusing'). If the temperature falls below the minimum temperature at which this free diffusion is normal (i.e., the melting temperature), the atoms will try to solidify; but before solidification can begin, a nucleus must be formed. A nucleus may be formed in one of the following two ways. It can be formed by the simultaneous arrival by chance of a sufficient number of atoms in the perfect geometric relationship for the formation of the minimum stable group of the basic crystal structure, capable of existing as a solid in the particular conditions. Alternatively, it can be formed by a sufficient number of atoms suitably grouping under some outside influence so that they become stabilised. The farther the temperature falls below the melting temperature, the stronger is the tendency to solidify. This tendency develops because atomic movement slows down progressively; giving (a) an increasing chance of a stable group forming, and (b) stability with a decreasing size of group. That is, the 'free energy' between the liquid and the solid state becomes greater.

Nuclei do not form readily. However, nucleus formation can be aided by outside influences, such as turbulence which might be caused by irregularities on the mould walls (a mould is a suitably shaped container inside which liquid metal is deliberately solidified), or local conditions in the vicinity of foreign atoms, or contained non-metallic impurities. Generally, stable groups will form in several different places at about the same time and solidification will proceed outwards from each, particularly in the directions of any thermal gradients set up in the liquid as heat is conducted away. Growth continues simultaneously from each nucleus and similar growth will also take place from any other nuclei that form subsequently, on the lines shown in Fig. 2.3. Separate nuclei are unlikely to be oriented on the same direction axes as each other; consequently, when two growing masses meet, they will not match one another, but will join up along a 'grain boundary', each forming a 'grain' bonded to the other by a combination of mechanical keying and a somewhat irregular metallic bonding. Although the average strength of atomic bonding along a boundary will not be nearly as great as in the body of a crystal, the atomically irregular nature and 'keying' of a boundary can prevent weakness being as great as might be expected. Grain size has great influence and shape some influence on the mechanical strength and plasticity of a metal. Usually the larger the grain size the lower the mechanical strength and the greater the plasticity, although grain boundary weakness in deformation may offset the latter effect. The effect of crystal shape is considered in Section 2.1.2.

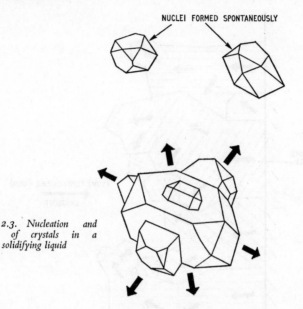

NUCLEI FORMED SPONTANEOUSLY

Fig. 2.3. *Nucleation and growth of crystals in a solidifying liquid*

Once solidification of an undercooled liquid begins, its temperature tends to rise back towards the melting temperature as energy is released from the solidifying metal, a phenomenon known as 'recalescence' (see Section 2.4.2).

2.1.2 Crystal Shape and Size in a Solidified Metal

Growth from a nucleus in a solidifying metal tends to proceed in the direction of the steepest thermal gradient, that is, against the direction of maximum cooling through the solid. However, atoms will attach more readily to some crystal faces than to others, thus, these faces may grow rapidly, even though they do not lie normal to the gradient, until other branching growth arms or 'dendrites' form more readily from developing faces in other directions (Fig. 2.4). The average main growth is always in the direction of the main thermal gradient.

The combined effect of nucleation of dendrites at a mould wall, and of a temperature gradient set up by cooling from the mould wall, is to encourage the various crystals to grow parallel to each other in 'columnar' form as shown in Fig. 2.5. Intersecting tips from opposed

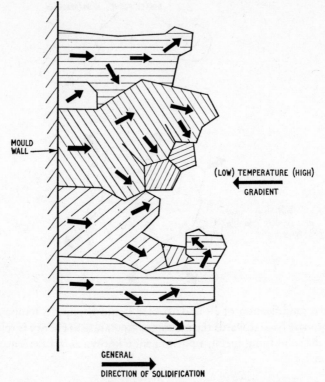

MOULD
WALL

(LOW) TEMPERATURE (HIGH)

GRADIENT

GENERAL
DIRECTION OF SOLIDIFICATION

Fig. 2.4. Dendritic growth tending to average out against the direction of
heat extraction from a mould wall

growth of such crystals tend to form planes of weakness (see Fig. 2.5)
both by entrapping impurities and by weaker bonding.

If the thermal gradient can be made steep enough to cause very
severe undercooling, then nuclei may be created more rapidly than
growth can take place and a large number of small equi-axial crystals,
called 'chill' crystals, may be formed.   Thin sections may solidify right
through as chill crystal structures.   However, chill crystals are likely
to occur only in the early stages of solidification of a large mass, even
when the latter is cooling in a water-cooled metal chill-mould; giving,
perhaps, an initial layer of fine equi-axial crystals superseded by normal
coarse columnar crystals as solidification proceeds inwards and the
temperature gradient decreases (see Fig. 2.6).   An 'equi-axial' crystal is

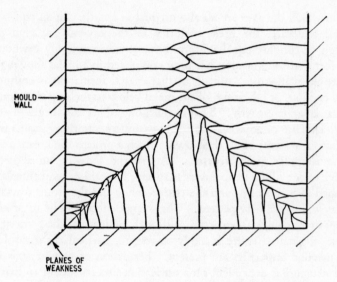

MOULD
WALL →

PLANES OF
WEAKNESS

Fig. 2.5. Columnar growth of a pure metal solidifying in a mould

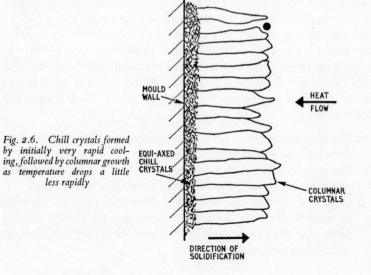

Fig. 2.6. Chill crystals formed by initially very rapid cooling, followed by columnar growth as temperature drops a little less rapidly

MOULD
WALL

EQUI-AXED
CHILL
CRYSTALS

HEAT
FLOW

COLUMNAR
CRYSTALS

DIRECTION OF
SOLIDIFICATION

one in which the axes are all about equal in length. In an equi-axial crystal structure, the great majority of the crystals are equi-axial crystals of roughly equal uniform size, usually randomly orientated. The fact that solidification has been completed in a metal showing no allotropic change may not mean that crystal formation is complete. Slow cooling, or reheating, of the metal below its melting temperature can cause 'grain growth'. That is, the most stable grains begin to grow at the expense of adjacent less stable grains by absorbing atoms from the latters' surfaces on to their own. Such a process can create a very coarse irregular grain structure. A material showing an allotropic change under falling temperature may have its grain structure markedly changed as it cools through the critical range and the atoms rearrange themselves to the new pattern. The overall effect may be to produce a finer more equi-axial structure, although evidence of the form of the coarse original structure is likely to remain, particularly if insoluble non-metallic impurities are present. Formation of new crystals in a solid structure is dependent on a nucleation process similar to that for nucleation from a liquid.

2.1.3 *Mechanical Breakdown of a Crystal Structure*

If a system of increasing mechanical load is applied to a polycrystalline mass of metal, the crystals will try to adapt their shapes to the various stress patterns that develop in the metal (i.e., they will 'strain') so that these internal forces can be resisted and equilibrium maintained. In the early stages, the pattern of strain will be almost entirely 'elastic' and, if the load is removed, the crystals will return to their original shape; but, if the load continues to increase, one of three things can happen:

1. The mass can begin to rupture by splitting along the grain boundaries, giving intercrystalline fracture.
2. The mass can begin to rupture by separation between adjacent planes in individual crystals, giving transcrystalline fracture.
3. Individual crystals can begin to deform plastically by shearing movement between adjacent crystallographic planes; usually those with the closest packing of atoms, since they are the most widely separated planes.

Which event will occur depends on the relative strengths and orientations of grain boundaries and crystallographic planes and on the type of loading. Plastic slip takes place only if shear stress is present; therefore, such deformation is impossible under either hydrostatic pressure or equilateral triaxial tension. Usually, shear stress is present and most metals are capable of undergoing plastic deformation so the most likely sequence is elastic strain, followed by further elastic strain and increasing amounts of plastic strain as stress intensity rises. Continued increase in loading leads to progressively greater amounts of plastic strain and the elastic component of strain tends to become relatively unimportant.

As plastic slip takes place in any one crystal, the latter's changing shape must inevitably react on adjacent crystals which will react in their turn. Such interaction leads to a change in stress distribution within the material that, in its turn, causes changes in the plastic slip pattern, bringing more, differently oriented, crystallographic planes into the movement. Gradually, a turbulent flow pattern develops, grain boundaries begin to disintegrate and the basic crystal structure begins to lose its regularity. Eventually, if plastic flow continues, the poly-crystalline structure seems to lose its character, as shown in Plate 2.1. Intense residual stresses are left in a plastically strained metal when the load is removed. Breakdown of a metal structure in this way may be deliberately caused, to give a required shape or to develop certain desirable mechanical properties, in the process known as 'cold working'. The resultant effect of plastic flow on the mechanical properties is called 'strain hardening', because the hardness of the metal rises with the amount of plastic strain. At any stage in plastic deformation, critical grain boundary or transcrystalline stresses may be developed and local cracks may appear or total rupture may be caused. For more detailed discussion of these effects, see *Mechanical Properties of Metals* by the present author (Allen & Unwin).

2.1.4 *Stress Relief and Recrystallisation of a Deformed Metal*

If a plastically strained metal is heated, the atoms become more mobile. At slightly elevated temperatures the effect of atomic mobility is to relieve some of the residual stress. The effect increases with rise in temperature until a level is reached at which such residual stress is almost completely relieved. Further rise in temperature makes the atoms able to diffuse much more easily about themselves and as a result the basic structure begins to reform, or 'recrystallise', around the nuclei. Certain parts of the structure will be relatively more stable than others and so will be able to act as nuclei for the new crystal structure. These centres will grow at the expense of adjacent less stable parts of the structure by absorbing atoms from the latter. If suitable conditions are maintained, the whole structure will reform around the new grains. The number and therefore size of new grains will be determined, mainly, by the amount of prior deformation, the greater the deformation the greater the number of grains that will start to form simultaneously. Thus, a slight deformation, followed by reheating into the recrystallisation range, will produce a large grain size, and a large amount of deformation, initially, a small grain size.

Deliberate heating to reform the grain structure of a previously deformed metal is known as 'annealing', although the term is sometimes more loosely used also for certain stress relieving treatments (see Section 2.5.4).

If deformation of a metal is performed at a sufficiently elevated temperature, recrystallisation occurs simultaneously with the breakdown of the structure and the metal is said to be 'hot worked'. In this case, the forces required to cause deformation are likely to be greatly reduced, because ease of deformation tends to increase with rise in temperature and rate of recrystallisation.

In either annealing or hot working, if the temperature is raised excessively far above the minimum for recrystallisation or the material is maintained for too long a time at the elevated temperature, then, once the structure has reformed, grain growth will tend to occur. That is, the atom-absorbing process will continue with the larger more stable grains growing at the expense of the relatively smaller less stable grains until most or all of the latter shrink and disappear. This effect, if allowed to proceed too far, can seriously weaken a metal.

With many metals, the only method for refining the grain structure is by deformation and recrystallisation, either by cold working and annealing or by hot working. In either case, heating time and temperature must be controlled to avoid the extremes of insufficient recrystallisation and excessive grain growth.

2.1.5   *Fibre Structure*

In the course of plastic deformation the grains tend to elongate in the main direction of plastic flow, and in some cases may also tend to align in one direction of orientation. Such changes can impart varying mechanical properties related to the direction flow. The metal is said to be developing a 'fibre' structure. Such effects can be greatly exaggerated in the presence of impurities, either non-metallic or metallic. Impurities of a brittle type may tend to break up with severe plastic flow and at the same time spread out in the main direction of flow, see Plate 2.1. Directional distribution of this kind may also cause relative differences in mechanical properties. Because this latter type of fibre persists even if the parent structure is completely reformed by recrystallisation, it can help to indicate the nature of the deformation that caused it, if the structure of the metal is revealed by subsequent sectioning and suitable preparation such as polishing and etching.

In certain materials, recrystallisation may take place preferentially in certain directions relative to the initial deformation and the crystals will then tend to be oriented with a majority lying in, or nearly in, the same direction. An orientation of crystals in this way is known as 'preferred orientation'. Its effect can modify or accentuate any variation in mechanical properties that may be caused by other fibre effects. It should be noted that the direction of this orientation is likely to differ from that of any prior alignment caused by the plastic flow.

2.2   SOLUBILITY AND ALLOYING

It is well known that certain liquids will go into solution in each other, that is, that the atoms of one will diffuse into intimate mixture with the atoms of the other. It is also well known that certain solids will dissolve into certain liquids in quantities up to some limiting value. However, it is not so well known that certain solids can show the same tendencies towards each other and can show full or partial solubility, without any melting. Metals can show all these solution phenomena in differing degrees and the mechanism can be used to produce 'alloys'. Alloys are deliberately controlled combinations of a metal of one kind with quantities of other metals and/or non-metals, developed to improve the physical or chemical properties of the parent (or basic) metal, or to impart to it properties that it does not inherently possess. Basically there are two types of solution:

1. *Substitutional solid solution.* In this type of solution atoms of the solute displace atoms of the solvent, one for one, usually without causing any change of lattice pattern, although inevitably some lattice distortion is caused since two dissimilar atoms cannot have exactly the same atomic sizes and valencies. Usually, for this type of solution to take place over an appreciable range of composition, the atomic sizes must be roughly equal to each other and the valencies should not be too dissimilar. Suitable constituents can be completely soluble in each other only if their inherent lattice structures are also similar.

2. *Interstitial solid solution.* If the atom sizes of two elements are widely different it may be possible for the smaller sized atoms to penetrate into the spaces between atom positions in a lattice of the larger sized atoms. This kind of solubility causes severe distortion of the parent lattice and is likely to be limited to small proportions of the smaller atoms. The elements most likely to form solutes of this kind are Boron, Carbon, Hydrogen, Nitrogen and Oxygen.

Solid solubility may show marked changes with temperature and may be associated with more positive changes such as compound formation or structural changes such as allotropic changes.

Certain conditions are transition states that persist only because diffusion is so slow in the given circumstances that it prevents the structure rapidly attaining a more stable condition. Such a condition, or an alloy in that condition, is said to be 'metastable'.

The constituents in a given alloy may exist in various forms, ranging from complete indistinguishable solution to clearly defined 'islands' of one metal embedded in a matrix of another. Each distinguishable constituent or condition is known as a 'phase' and is usually represented by a Greek letter symbol (e.g., γ phase in steel).

It is by controlling the relative amounts and distribution of these phases that the metallurgist controls the properties of an alloy.

It is not only the mechanical properties that are influenced by alloying, but both stable and metastable conditions can influence other physical and chemical properties such as electrical conductivity and corrosion resistance. The main effect of some alloy additions is to aid nucleation and to minimise the grain size of the basic constituent of the alloy, thus improving its strength in the cast condition.

2.2.1 *Changing Solubility*

The most useful phenomenon to the metallurgist in controlling the condition of an alloy is change in solubility. Many metals are almost insoluble in each other in the solid state but are fully soluble in the liquid state; hence, by melting them together and then cooling them at a controlled rate, it is possible to produce a solid mixture of islands of a phase based on one metal embedded in a matrix of a phase based on another metal. Controlling the rate of cooling can often control both the fineness of the phase distribution and the geometry of dispersion. The main governing factor is the speed with which the constituents tend to separate during the freezing and subsequent cooling process, relative to the stage at which the solid mixture becomes so interlocked and rigid that diffusion becomes too slow to have a significant effect. The relative amounts of different phases are governed mainly by the particular composition of an alloy.

Dispersion of a constituent can be even more effectively controlled if it is fully soluble in the solid state at an elevated temperature and relatively insoluble at lower temperatures. If the solid mixture is heated to the appropriate elevated temperature and held there for a sufficient length of time, it will form one solid solution phase with the solute atoms dispersed more or less separately and uniformly throughout the solvent matrix. Rapid cooling from the elevated temperature may then entrap the solute atoms and produce a 'supersaturated' solid solution. If diffusion is sufficiently rapid during cooling, the solute atoms will first cluster to form stable nuclei, then 'precipitate' by growth of nuclei to form optically distinguishable phase groups, the largest groups usually being at the grain boundaries of the solvent. Usually, as the available diffusion time increases the number of the individual particles of precipitated phase likely to be formed, decreases, and their average size increases.

Certain critical sizes and degrees of dispersion of precipitated particles give optimum effects on the properties of the alloy, in particular, the mechanical properties, and control of the latter by this means is known as 'precipitation hardening'. According to the materials and the desired results, critical conditions vary from atomic scale uniform dispersion of the retained constituent in the solvent, through fine distribution of the constituent along grain boundaries of the solvent, up to formation of microscopically visible precipitated particles as shown in Fig. 2.7. Mechanically, the most common effects of such conditions are (*a*) to raise the stress required to initiate plastic flow in the material, that is to raise its yield or proof stress, and (*b*) to increase the rate of work hardening during plastic flow, which will increase the total load carrying capacity or nominal breaking strength of the alloy.

Allotropic changes tend to give dramatic changes in solubility. Sometimes full solubility at an elevated temperature is exchanged for almost complete insolubility at only a slightly reduced temperature in the presence of an allotropic change. Naturally, the metallurgist tries to make full use of such an effect and it forms the basis for the quench hardening and tempering of steel, see Section 2.2.5; but it must be borne in mind that for any of these phenomena to be usable, the results must be sufficiently stable to show no significant change in properties with time at service temperature.

2.2.2 *Precipitation and Ageing*

If a solution is supersaturated to any degree, the solute will always be diffusing out of the solution as it tries to reach more stable conditions. The rate at which the diffusion takes place depends largely on the relative ambient temperature, therefore, the time taken to achieve stability can vary according to temperature. Since a state of stability generally implies reducing an alloy to a low level of apparent strength, it is essential, if such an alloy is to be used in a precipitation hardened condition, that the rate of diffusion should be so low that no significant change occurs during its service life. Change of strength with time is called 'ageing'—'natural' ageing if it occurs at room temperature and 'artificial' ageing if elevated temperature is required.

Fig. 2.8 shows a typical form of curve illustrating the time taken to reach the stage of complete precipitation at differing temperatures. Near the temperature of solution, the time is long because nucleation is difficult. Time is also long at low temperatures because the diffusion

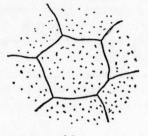
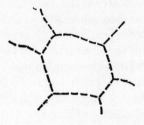

(a)

(b)

(c)

Fig. 2.7. *Forms of distribution of an alloy phase in a matrix.* (a) *Atomic scale dispersion.* (b) *Atomic scale grain boundary precipitation.* (c) *Early stages of clustering*

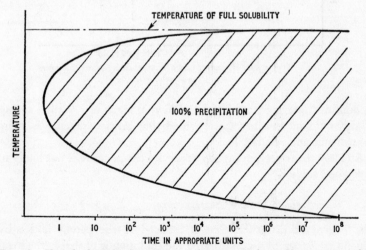

TEMPERATURE OF FULL SOLUBILITY

100% PRECIPITATION

TEMPERATURE

TIME IN APPROPRIATE UNITS

Fig. 2.8. *Typical relationship between time and temperature for 100% precipitation of the solute from a supersaturated alloy during ageing*

rate is very low. Between the relatively lower and higher temperatures there is a minimum time which may be only a few seconds. The
relative length of the longer times depends on the alloy and may vary
from a short time of a few hours to a period of hundreds of years. For
an alloy to be useful in the artificially hardened condition, the time for
complete precipitation at the service temperature(s) must be considerably longer than the expected service life if no serious change in
properties is to occur.

If optimum strengthening does not occur with maximum dispersion
of the solute in a supersaturated alloy, it may be possible to use ageing
for strengthening the alloy by 'age hardening' it until the critical

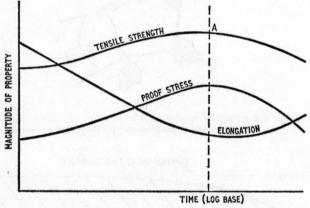

*Fig. 2.9. Typical form of property variation curves during ageing of a super-
saturated solid solution*

amount of clustering is achieved as shown in Fig. 2.9. Natural ageing
of this kind in an alloy can be used, but artificial ageing is more
controllable and reliable.

An alloy that overages rapidly at service temperature is not likely to
give good service.

2.2.3   *Strengthening a Lattice by Supersaturation*

The degree of distortion that is induced in the parent lattice by
generating a state of supersaturated solid solution is probably the main
factor that affects the mechanical properties of the material.

Most metals are weak because they are plastic and they are plastic because certain crystallographic planes will slide easily across each other with the help of 'dislocations'. Dislocations are certain inherent defects that are present in the metal and that can be generated under stress. Anything that inhibits plastic sliding will raise the yield stress of the metal and will also raise its apparent strength by raising the rate of work hardening and preventing the development of localised attenuation of cross-section ('necking'). The simplest way to reduce plastic sliding is to distort the crystallographic planes, and so stiffen the lattice. This reduction is achieved not only because actual slip in an isolated crystal is made more difficult, but because the distortion makes it much more difficult to propagate slip from one crystal to another in a polycrystalline mass. A fine grain size accentuates the latter effect.

Stiffening of a lattice by alloying can be achieved either by forming a stable solid-solution alloy of the basic metal or by forming a super-saturated solid-solution alloy. In either case, the strengthening is pro-portionate to the amount of solute in solution, but more spectacular results are likely with smaller amounts of alloying addition if super-saturation is used, since the presence of alloying atoms in the parent lattice, in positions in which they would not normally be tolerated, causes severe lattice distortion; the less the tolerance the greater the distortion. In some materials, it is atomic scale distribution of a retained solute that gives optimum distortion, in others it is some degree of clustering; therefore, the treatment given to alloys to attain improved strength varies with the constituent materials.

In every case, the critical degree of dispersion of alloying elements must be both readily achieved and steadily maintained if the alloy is to be useful. Therefore, where the desired result of alloying is control of mechanical properties, any necessary heat treatment should be applicable entirely in the solid state, see Section 2.2.1, rather than by cooling from liquid to solid. This usually means heat treatment hardening by one of two systems: (a) solution treatment and ageing or (b) quench hardening and tempering.

2.2.4 Solution Treatment and Ageing

If the main constituents of an alloy are completely soluble in each other in the solid state at an elevated temperature and decrease rapidly in solubility with decreasing temperature, then the alloy may be capable of showing improved mechanical properties by a three-stage process of (a) solution treatment, (b) quenching and (c) ageing.

Initially, the alloy is heated to as high a temperature in the solid state as is reasonably safe without causing undesirable reactions in the material (e.g., oxidation, grain coarsening or incipient melting), and maintained at temperature until full solution has been attained. The time taken for the latter will depend on the nature of the constituents and on the relative temperature level. Higher temperatures tend to give more rapid diffusion and shorter times.

The second stage consists in artificially cooling, or 'quenching', the alloy so quickly that it is retained in a supersaturated solid-solution condition.

Final treatment takes the form of ageing, either at room temperature, if the alloy can be aged to the right degree by that means, or at a suitable elevated temperature. The time taken to age to optimum conditions depends on the degree of metastability relative to the treatment temperature. Subsequent usefulness of the treated alloy depends on its metastability at service temperature. Most industrial alloys, capable of responding to this treatment, have to be artificially aged because alloying additions are made deliberately to prevent natural ageing, since the latter could be dangerous if it led to overageing and deterioration of the properties.

Alloys of many metals can be effectively solution treated and aged, but generally, the properties are not spectacularly improved. Aluminium alloys of the 'Dural' type are probably the best known solution-treated and aged materials, the proof strength and tensile strength of many of these alloys may be improved by a factor of 2–3 over those of the same materials in the annealed condition.

2.2.5   Quench Hardening and Tempering

As mentioned in Section 2.2.1, solid solubility of one metal in another may change abruptly with a slight fall in temperature in the presence of an allotropic change. In this case, with the appropriate fall in temperature, solute atoms may become almost completely incompatible with the parent lattice and their presence in the highly supersaturated condition can induce a maximum drastic stiffening of the lattice.   Thus, if quenching can be rapid enough to give a state of maximum supersaturation and the metastability is great enough to retain this state, then quenching such an alloy from the condition of full solution will immediately give maximum strength.   However, this state is usually not desirable, because it is likely to be accompanied by excessive brittleness and sensitivity to shock, owing to the almost complete inhibition of plastic flow.   In some cases, cracking may be caused even by the quenching process.   The degree of lattice stiffening of a quench-hardened alloy may have to be reduced to make the material safe for subsequent use.   This is done by 'tempering' the material.   That is, its temperature is raised to a level where the seemingly suspended process of precipitation can begin to operate and the lattice can become progressively less rigid with time at the treatment temperature.   Such treatment usually reduces the proof and tensile strengths of the material, while increasing its plasticity and notch toughness (see Fig. 2.10).   Treatment is applied long enough to give the desired combination of properties.   Thus, quench hardening is usually a three-stage treatment similar in principle to solution treatment and ageing, but is described in different terms because of the effect of the change in structure.

Iron is a metal that shows an allotropic change and simultaneously an abrupt change in its solvency, particularly for carbon.   When alloyed with carbon to form 'steel' it is almost the only common engineering material suitable for quench hardening and tempering. This position it holds mainly as a result of its comparative cheapness. Because of its importance it is dealt with in more detail in Section 2.5.

2.3   REPRESENTATION OF PHASE CHANGE IN METALS

When the range of possible mixtures, compounds, solutions, etc., that can be formed by different combinations of even a few metals is considered, it will be realised that some means must be devised for giving as simple a guide as possible to the likely state of a given alloy composition at a given temperature.   Such a guide is essential for a worker in metals if he is to be able to practise his art.   It is impossible to devise one system that will give all the answers, even for a very simple alloy,

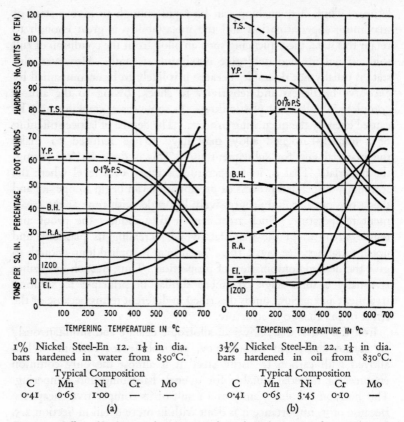

Fig. 2.10. *Effects of hardening and tempering on the mechanical properties of two similar steels containing different amounts of nickel.* (After The International Nickel Company (Mond) Ltd.)

1% Nickel Steel-En 12. 1⅛ in. dia. bars hardened in water from 850°C.

Typical Composition

C	Mn	Ni	Cr	Mo
0·41	0·65	1·00	—	—

(a)

3½% Nickel Steel-En 22. 1¼ in. dia. bars hardened in oil from 830°C.

Typical Composition

C	Mn	Ni	Cr	Mo
0·41	0·65	3·45	0·10	—

(b)

but quite a lot may be done with the aid of a 'phase equilibrium diagram'. A phase equilibrium diagram is drawn two-dimensionally for a binary (two constituents) alloy but a three-dimensional model must be used for complete representation of a ternary or quaternary (three or four constituents respectively) alloy system. Use of three-dimensional diagrams is complex and best left to the expert. However, the binary systems are relatively simple to understand and can, with discretion, be generally used.

With binary diagrams, the convention is to represent the proportions of the two constituents on the abscissa between two vertical parallel ordinates and on the ordinate scale to represent temperature, see Fig. 2.11. Between the ordinates, lines can be plotted to represent boundaries between phase conditions. As shown in Fig. 2.11, one such line, the 'liquidus', represents the limits of the completely liquid state

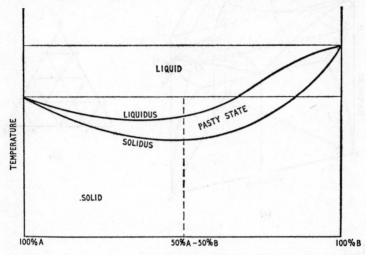

Fig. 2.11. *Two metals (A and B) showing complete solubility in each other, but tending also to show a eutectic*

and another, the 'solidus', represents the limits of the completely solid state. Between these two is represented a pasty state, which is neither completely liquid nor completely solid but is a mixture of liquid and solid. Changes in the solid state are marked by similar boundaries and particular solid state phases are usually designated by Greek letter symbols.

It must always be kept in mind that a phase equilibrium diagram represents *equilibrium* conditions attained by very slow cooling or heating of an alloy to the relevant temperature and holding it there until the alloy structure is as stable as it can be expected to be. The diagrams do *not* necessarily represent the conditions attained if an alloy is more rapidly heated or cooled.

Ternary diagrams are drawn with three equally spaced ordinates representing the three constituents, on the lines shown in Fig. 2.12 (a) and (b), and phase boundaries are marked by surfaces instead of lines; but this kind of diagram is not considered further here.

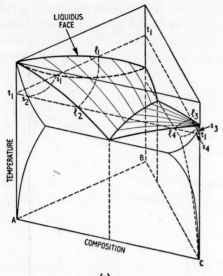

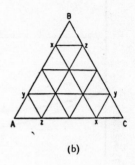

(b)

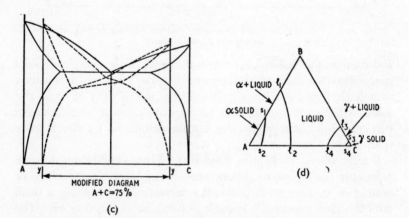

Fig. 2.12. *Ternary equilibrium diagram. (a) Imaginary diagram for metals A, B and C. (b) Plan showing lines of constant proportion of individual elements, e.g., xx is 25% and yz 75% of A. (c) Relationship between A and C with B constant at 25% (d) An isothermal 'section' at temperature t_1*

Understanding of binary systems is helpful when interpreting the behaviour of many alloy systems, so the following sections give a brief outline of the method of representation of typical alloy reactions. In the examples, the diagrams are all related to copper alloys to show the variety of possibilities that can be found, even in the presence of one common constituent. In every case, an ordinate dropped to the point representing the composition of a given alloy will pass through all the *stable* phase change boundaries to be expected with slow change in temperature for that particular alloy.

2.3.1 *Full Solubility*

As suggested in Section 2.2, if two metals have the same type of basic lattice and roughly equal atomic size and valency then they may be completely soluble in each other in all proportions. These conditions exist for copper and nickel and are represented in Fig. 2.13.

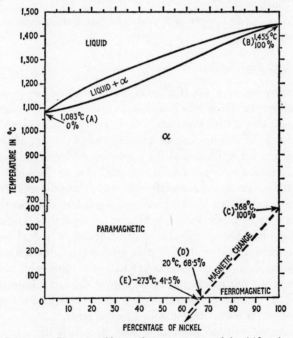

Fig. 2.13. *Binary equilibrium diagram: copper–nickel. (After the Copper Development Association)*

Only two lines (a solidus and a liquidus) and one symbol to represent the solid phase are needed to complete the diagram, since the only changes in the solid state are changes in magnetic characteristics with high proportions of nickel, occurring at appropriate temperatures, which need not necessarily be shown on the diagram.

With metals having less favourable characteristics, the liquidus and solidus lines would not be so symmetrical relative to each other and might be more like those for the imaginary system shown in Fig. 2.11. The range of solubility might also tend to be less and changes in solid state become more likely.

2.3.2 Virtually Complete Insolubility

Some metals, fortunately not many common ones, are almost completely insoluble in each other in both the liquid and the solid state, making them very difficult, if not impossible, to fabricate into useful alloys. Such conditions, for copper and molybdenum, are indicated in Fig. 2.14. The copper and molybdenum always exist as separate constituents.

2.3.3 Liquid Solubility–Solid Insolubility: Eutectic

Intermediate between the two previous types of relationship, there is the condition in which two constituents are completely soluble in the liquid state and almost completely insoluble in the solid state. Two typical forms of diagram can be encountered in these conditions.

In the first form, illustrated for copper and bismuth in Fig. 2.15, a degree of solubility is possible in the solid part of the pasty state that exists between the liquidus and the solidus. Thus, the solid part of the pasty state is designated α, but, below the solidus, which in this case is constant at 270·3°C, the basic constituents are separate. The minimum of the liquidus temperature almost coincides with 100% bismuth and with the melting temperature of pure bismuth, i.e., there is no marked intermediate droop in the solidus line to a minimum value below both the parent melting points. A minimum of the latter kind is, strictly speaking, a 'eutectic point' although a more specialised meaning, discussed below, is more often attached to this term.

The second form of diagram is that in which there is a definite 'eutectic'. A eutectic is generally looked upon, not only as a minimum

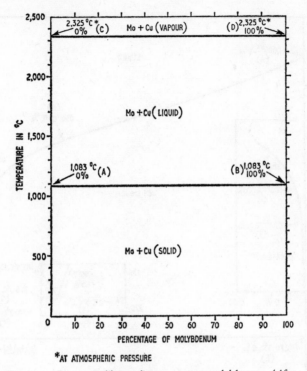

Fig. 2.14.   Binary equilibrium diagram: copper–molybdenum.   (After
the Copper Development Association)

liquidus temperature of an alloy system, but as that particular type of
minimum liquidus temperature at which the pasty state disappears and
solidification takes the form of a simultaneous solidification of the two
constituents, without fall in temperature until solidification is complete.
This occurs at point D in the copper–silver system (Fig. 2.16), and the
resultant eutectic structure in this case is illustrated in Plate 2.2.   Since
there is a slight solubility of silver in copper and vice versa, the solid
phases are referred to as α and β respectively, and the solid at room
temperature is, therefore, a solid mixture of these α and β phases.   An
alloy system containing such a eutectic is characterised by the marked
discontinuity in the liquidus line and the disappearance of the pasty
state at the eutectic temperature by final solidification of eutectic
composition liquid to form a eutectic composition matrix around

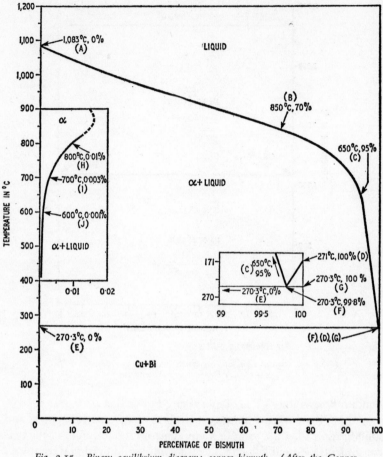

Fig. 2.15. Binary equilibrium diagram: copper–bismuth. (After the Copper Development Association)

grains of the constituent that is present in excess of eutectic composition (see Plate 2.2).

2.3.4 *Peritectic Reaction*

Another important reaction is the 'peritectic' reaction. This takes place between the solid and liquid metal in the pasty state at a fixed temperature and composition during cooling, and produces complete

solidification by formation of a new single phase of a fixed composition intermediate between those of the initial solid metal and the
liquid. The shape of the diagram in this area looks rather like an
inversion of a eutectic diagram and is illustrated in Fig. 2.17 which
shows part of the diagram for copper and tin (for up to 32% tin).
The peritectic is at C where cooling liquid and α solid react at 798°C to
form β solid.

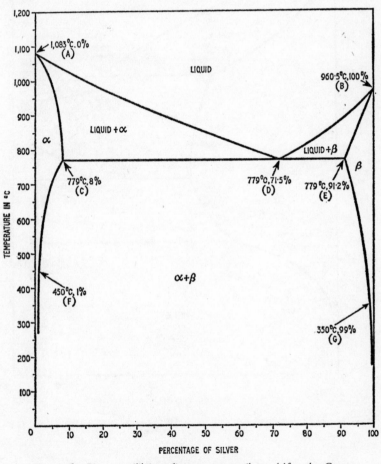

Fig. 2.16. Binary equilibrium diagram: copper–silver. (After the Copper
Development Association)

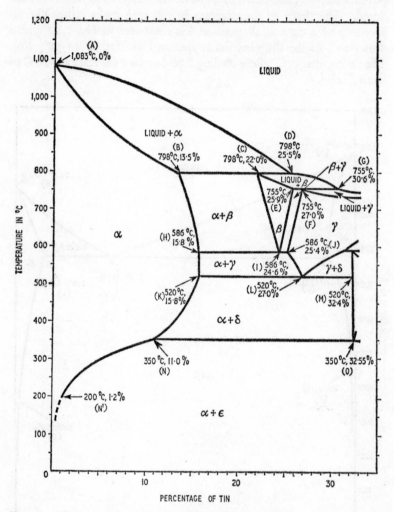

Fig. 2.17. *Binary equilibrium diagram: copper–tin (up to 32% tin). (After the Copper Development Association)*

2.3.5   *Compound Formation*

Certain proportions of particular constituents form compounds. Such compounds behave as individual constituents and react independently with any remaining amount of either one of the original constituents present over and above the proportion required to form the compound.   This behaviour has the apparent effect of breaking one complete diagram into distinct bands numbering one more than the number of compounds formed.   Each band has the character of a complete diagram as can be seen in Fig. 2.18 for copper and magnesium. Compounds are formed at compositions *A* and *B*, and each of the three bands has its own eutectic in this particular system.

2.3.6   *Eutectoid Reaction*

When a reaction, similar to a eutectic reaction, occurs entirely in the solid state and one solid solution phase of two constituents transforms, at a constant temperature during cooling, simultaneously to form two separate phases, it is called a 'eutectoid' reaction.   Two such reactions can be seen at *I* and *L* respectively in Fig. 2.17.   The appearance of a eutectoid structure is usually quite like that of a eutectic, but very much finer (e.g., see Plate 2.3).

2.3.7   *Peritectoid Reaction*

If a mixture of two solid phases transforms at constant temperature, during cooling, to form one third solid phase of intermediate composition, this is called a 'peritectoid' reaction since it is similar to a peritectic reaction.   This reaction, although rather uncommon, is illustrated in Fig. 2.19, for copper and silicon, at point *A*, where α and β transform to κ at 842°C, on cooling.

2.3.8   *Typical Alloy Systems*

Even from the few diagrams that are shown in Figs. 2.13–2.19, it will be clear that few of the reactions that have been described occur in isolation, even in any one binary alloy system.   In fact, nearly every complete diagram is likely to show several reactions each occurring, of course, at different compositions and temperatures.   By following the ordinate of composition up or down on a diagram, the equilibrium

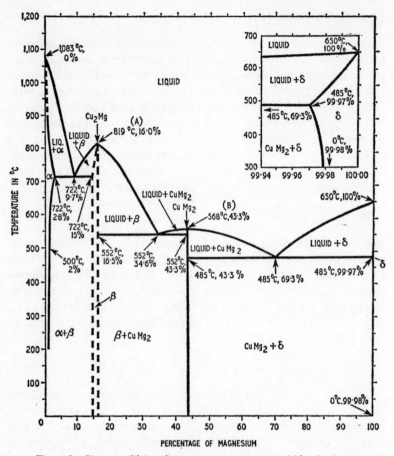

Fig. 2.18. Binary equilibrium diagram: copper–magnesium. (After the Copper Development Association)

heating or cooling (including melting or freezing) behaviour of a given alloy may be studied. In the presence of a simple combination of reactions in a system, backed by some experimental verification, it may also be possible to predict non-equilibrium behaviour of certain alloys; but, as complexity increases, it becomes progressively more difficult. Naturally, the complexity of alloys containing more than two constituents tends to be greater than with two alone, hence, ease of interpretation decreases.

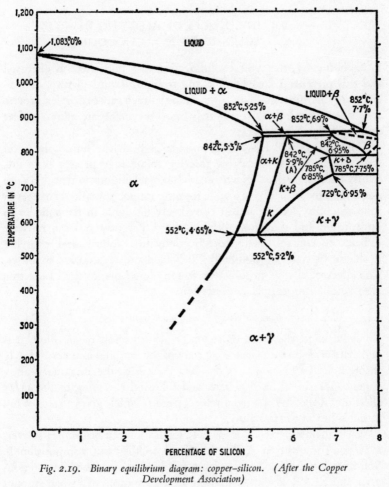

Fig. 2.19. Binary equilibrium diagram: copper–silicon. (After the Copper Development Association)

Nevertheless, equilibrium diagrams can prove exceedingly useful in studying solidification and bonding characteristics of particular alloys in relation to both casting and welding.

2.4 INFLUENCES OF ALLOYING IN SOLIDIFICATION AND COOLING

In Section 2.1.1, the typical solidification of a pure metal is discussed and this presents a fairly simple and straightforward picture. However, the solidification of an alloy can be a much more complex process and, since most industrial materials based on metals are alloys, a clear understanding is necessary.

Those alloys which are fully soluble in both the liquid and solid state and have a very narrow freezing range may present little difficulty and may behave in a manner closely analogous to that of a pure metal; but those that have a wide freezing range, or marked changes in solid solubility, or are almost completely insoluble in the solid or in both liquid and solid may behave very differently and can be very difficult to control. The three sections that follow deal with the problems of changing solubility, shrinkage and segregation in alloys. The effect of change in solubility is considered first as the easiest step after considering simple solidification.

2.4.1 *Alloy Composition and its Effect on Cast Structure*

To understand the main difficulty of solidification of an alloy, it is necessary to follow the changing state of the material as it cools slowly through its pasty range. Consider, therefore, the imaginary alloy system $X–Y$ shown in Fig. 2.20, and follow the cooling of the 55/45 alloy down the line AA from temperature t_1 which gives a state of full liquid solution for this alloy.

When the temperature drops to t_2, nucleation can begin. However, it is not composition A that begins to solidify but composition B. Solidification begins by the same mechanism described in Sections 2.1.1 and 2.1.2, but with separation or precipitation, of a small amount of solid of composition B. This entails a change in composition of the remaining liquid by enrichment in metal X, since the *average* composition of the alloy as a whole must remain unchanged. Fall in temperature towards t_3 leads to further precipitation of solid, but its composition

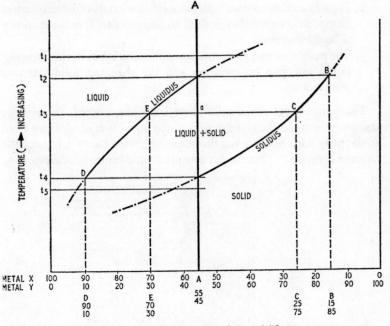

Fig. 2.20. *Composition variation during solidification*

changes progressively with decreasing temperature down the solidus
line towards composition *C*; the composition of the remaining liquid
will also change towards *E* to keep the average composition constant.
Thus, at temperature t_3 the pasty state will be a mixture of solid composition *C* with liquid of composition *E*.   The question remains, how
much of each?   The average must lie at *a*, so a simple 'lever' rule can
be applied, the arm *ca* being proportionate to the quantity by weight,
of liquid (*ca/CE*), and the arm *aE* being proportionate to the quantity
of solid (*aE/CE*), if *CE* represents 100% at temperature t_3.   This allows
the relative quantities by weight of solid and liquid to be calculated
for any temperature *under equilibrium conditions* of an alloy of
known composition.   To achieve equilibrium at t_3, three things are
necessary:

1. The initial solid must change in composition from *B* to composition *C* as the temperature decreases.

2. Liquid metal in contact with the solidification front interface must change in composition from A to composition E as the quantity of liquid decreases.

3. The pasty state must accommodate diffusion between the changing liquid interface composition and the changing solid interface composition.

The composition of the solid phase is tending towards the average composition while the composition of the liquid is tending away from it, the pasty stage maintaining the balance between the two, to keep the average constant. Thus, under complete equilibrium conditions at t_3,

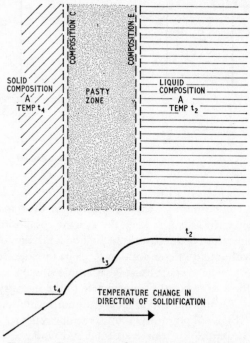

Fig. 2.21.   Transitory state of compositions and temperatures
at solidification front during normal cooling of composition A
(With reference to Fig. 2.20)

the alloy will be a mixture of solid of composition C and liquid of composition E; but, if a temperature gradient exists, the instantaneous condition of a solidification front might be as shown in Fig. 2.21.

A further drop in temperature, to t_4, continues the solidification and diffusion process and at this temperature the material becomes 100% solid just as its composition arrives at the average value A, whilst, just before this point is reached, the very last drop of liquid reaches composition D.

Below t_4 to say t_5, cooling proceeds with no change other than simple contraction. No compositional or structural change takes place because no phase change boundary is crossed. If a solid phase change boundary is crossed, then the relevant solid phase changes will take place in a manner similar to that described above, except that diffusion is likely to be very much slower.

In making the above statements in this section, it is assumed that equilibrium is maintained within each local volume of material as it reaches the temperature under consideration. That is, it is assumed that either the rate of cooling is infinitely slow or the rates of diffusion are infinitely fast. However, neither one of these conditions is likely to prevail in normal solidification, the cooling gradient is likely to be relatively steep and the rate of diffusion is likely to be limited and will certainly differ between liquid and solid. For these reasons, it is likely that composition gradients will be set up in the material (e.g., see Fig. 2.21), and equilibrium will not be maintained, particularly in solid phases.

In these circumstances, in the example considered above, alloy A (Fig. 2.20) cooling towards t_4 will begin to solidify at composition B and temperature t_2, but the solid phase will not have time to diffuse towards composition A. By the time the temperature reaches t_4, the solid will be locally impoverished and the liquid locally enriched, relative to constituent X. Solidification by growth slows down and the liquid begins to undercool below the nominal solidification temperature. Now, as the undercooling increases, the drive for spontaneous nucleation (Section 2.1.1) increases and the rate of diffusion decreases in the liquid so that a time may be reached when the remaining liquid metal finds it easier to nucleate spontaneously within itself rather than to precipitate on to the already solidified metal with its compositional gradient. In such circumstances, an alloy ingot that has been solidifying in a similar manner to a pure metal, as illustrated in Figs. 2.5 and 2.6, changes its growth mode and may complete solidification by forming an equi-axed crystal structure as shown in Fig. 2.22.

Such final equi-axed crystals are very much coarser than the equi-axed chill crystals that may be formed when solidification first begins (Section 2.1.2), because the frequency of generation of nuclei in the

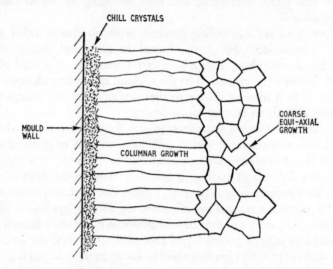

Fig. 2.22. *Changing macrostructure in an alloy caused by composition gradient becoming too steep for complete columnar growth*

liquid is lower. It should be noted that this kind of change of growth mode can occur only in an alloy.

Retarded diffusion, as described above, can have two effects. If a composition gradient is set up in a solid phase during solidification or on cooling through a range of changing solid solubility and the final state is metastable, then the composition of an individual grain or crystal may vary gradually from its nucleation centre outwards to the grain boundary. Such a grain is said to have a 'cored' structure. On the larger scale of solidification, if the composition of the first crystals to form during solidification differs so markedly from that of the last crystals to solidify that a large proportion of one constituent is found in one region and a significantly smaller proportion in another, perhaps with a gradual change in composition in intermediate material, the structure is said to be 'segregated'.

The physical, chemical and mechanical behaviours of a segregated material are likely to differ considerably from those of a homogeneous alloy and even slight coring can cause marked changes.   Because of its importance, segregation is considered in more detail in Section 2.4.3.

2.4.2   Shrinkage in Solidification; Tearing and Pipe

So far, little attention has been given to the influence of thermal shrinkage on the structure of a solidifying and cooling metal; but it is apparent that it must have important effects.

As a liquid metal cools it contracts at a slow rate, relative to temperature fall, then as solidification takes place there is a marked contraction followed by slow contraction of the cooling solid.   All common metals will go through this sequence although each has its own characteristic rates and mode of contraction.   If a solid metal cools through an allotropic change then shrinkage behaviour may be momentarily modified according to the nature of the allotropic change.   In cases where the allotropic change is to a state of less dense atomic packing, as in the change from face-centred cubic (γ) iron to body-centred (α) iron (see Section 2.2.5) the shrinkage will alter to expansion during the change.   Expansion continues until the change is complete, then contraction is resumed.   Such an allotropic change always involves the release of energy as heat, so the temperature fall is arrested during the change.   If cooling is rapid, the temperature may drop well below the temperature of the allotropic change before the change gets fully under way.   In such circumstances, the released energy from the reaction when it begins may be sufficient to raise the temperature back towards the stable change temperature, giving an effect called 'recalescence', see Fig. 2.23.   Similar effects occur during solid phase changes in a cooling alloy.

These phenomena are reversed during a heating cycle.   A structural change requires extra energy if it occurs during heating, so the temperature rise is arrested until the change has taken place.   Thermal contraction of a solid polycrystalline material has other effects, which vary with the characteristics and shapes of the individual grains. Most crystal structures contract more on some axes than on others, therefore, the grain will also contract anisotropically.   If the grains are equi-axial, then such uneven contraction is not likely to cause much variation in the overall uniformity of contraction of the polycrystalline mass, because the effect is averaged out.   The worst that is likely to

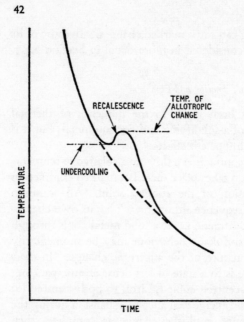

Fig. 2.23. *Recalescence during an allotropic change on a cooling solid*

happen is that some local stress concentration may be set up and a few microcracks be generated at some badly orientated grain boundaries. If the grains are large, elongated as in columnar growth or preferentially orientated coaxially, the contraction process may generate more serious effects.

In any polycrystalline mass, the individual grains will shrink about their own centres, that is to say, they will tend to draw away from each other at their boundaries in certain directions. At the same time, any remaining liquid will also be contracting rapidly. The drawing apart of crystals, and particularly the contraction of any pasty material bridging them, will set up tension across the boundary (or boundaries) where it occurs and will do one or more of three things; (*a*) leave residual stress and elastic distortion, (*b*) cause plastic distortion, or (*c*) cause grain boundary rupture or 'tearing' of the weak material between grains. All three effects can occur in varying degrees according to the characteristics of the material and amount of contraction. A material with weakly bonded grains or a brittle structure will develop an extensive network of microcracks or tears or, at the worst, may even disintegrate. An alloy with a wide freezing range is more liable to tearing than one with a narrow freezing range, although the fluidity of the last metal to solidify and the plasticity of the solid it forms will also influence the results.

During the solidification from the liquid state, the shrinkage of the solidifying liquid together with that of the individual solidified crystals create gaps into which some of the remaining liquid will be drawn, if it is sufficiently fluid. The activating power forcing the

PIPE

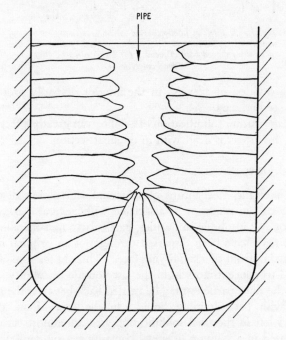

Fig. 2.24. Formation of a 'pipe' by draining into crevices between contracting crystals of last liquid to solidify

liquid in may be a combination of capillary attraction, fluid pressure and pressure from gas evolving from the liquid. The overall effect is that some liquid metal, from the last zones to solidify, drains away into the solid metal, leaving a cavity. In the case of an ingot in which a fairly large mass of metal is solidifying, disappearance of such liquid leaves a conical cavity, called a 'pipe', running down into the ingot as shown in Fig. 2.24. On the smaller scale in a casting, it can leave

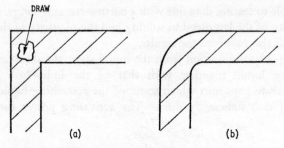

Fig. 2.25. *Effect of locally delayed solidification within a body of metal. (a) A local cavity called a 'draw' formed by delayed cooling of locally enlarged bulk of metal. (b) Draw cured by eliminating local enlargement*

shrinkage cavities, or 'draws', in the interior of badly proportioned sections, see Fig. 2.25.

Shrinkage during solidification of an alloy can greatly affect segregation and this aspect is dealt with in the next section.

2.4.3 *Segregation*

Operation of the mechanisms of structural segregation described in Section 2.4.1 can cause, or be associated with, several marked effects.

Most of any insoluble impurities, particularly non-metallic impurities, that may have been dispersed throughout an original liquid are transported in front of a solidifying face. Thus, in the case of a metal, they tend to concentrate within the last volume to solidify, although some of the finer particles may be carried back between the shrinking crystals with back-draining liquid. In the case of an alloy, the impurities left in the final solidification area are almost certain to be concentrated in a volume of metal containing a substantially greater proportion of the lower melting point constituent(s). This type of concentration of impurities and segregation of structures is often called 'major segregation' because it is usually the most obvious and widespread effect.

A somewhat similar effect, on a much smaller scale, can occur where two solidification fronts intersect each other at an angle, as shown in Fig. 2.26 (a) for the intersection of columnar growths at a sharp corner. In this case, some impurities and lower melting point constituents are likely to be entrapped in the interstices of the crystal tips along the line of intersection, although this volume may not be the last part of the main mass to freeze.   When it could cause serious weakness, this effect is minimised by rounding off external angles so that the growing crystals do not interlock at their tips, see Fig. 2.26 (b).

It is to be expected that the draining back of liquid metal into gaps between previously solidified shrinking crystals will affect segregation. The effect on impurities is mentioned above; but it will also affect structural segregation and give rise to 'inverse segregation' in an alloy. If some of the last liquid to solidify, perhaps driven by pressure set up

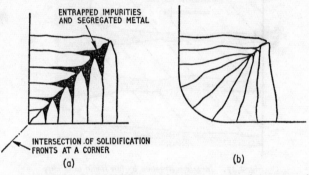

ENTRAPPED IMPURITIES
AND SEGREGATED METAL

INTERSECTION OF SOLIDIFICATION
FRONTS AT A CORNER
(a)   (b)

Fig. 2.26.   Local effect of columnar solidification; entrapping of impurities and segregated material at intersecting solidification fronts.   (a) Local segregation. (b) Minimising such segregation

by the evolution of gas, drains back between the first crystals to solidify then two dissimilar compositions may be found close beside each other without any marked graduation in composition between them, on the lines shown in Fig. 2.27.   In extreme cases, the liquid metal can drain right back to the outer surface, to appear as 'bleeding', and solidify as irregular beads, or 'blebs', on the material surface.   Such inverse segregation may be beneficial in reducing the average distance of segregation in the mass and in reducing the average variation in composition from place to place, making uniformity easier to attain subsequently, by a process of annealing diffusion or mechanical kneading or both.

For most alloys, there is an optimum rate of solidification cooling that gives the worst segregation conditions.   Infinitely slow cooling would probably give little segregation in a stable solid-solution alloy but would give very large grains and take an impossibly long time.

On the other hand, solidification at a rate so rapid that small equi-axial chill crystals, of fairly uniform composition, are obtained throughout the solidified mass is not always attainable and, in any case, many materials could not withstand the high contractional stresses that would

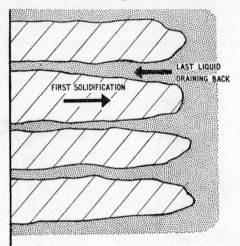

Fig. 2.27.   Inverse segregation by last liquid to solidify
(say composition D, Fig. 2.20) draining back past first
metal to solidify (say composition B, Fig. 2.20)

be set up.   As a rule, a compromise is accepted and the maximum acceptable cooling rate is used, perhaps followed by mechanical working (kneading), to blend the material, and/or heat treatment in the solid state.

2.4.4   Gas Evolution During Solidification

In several parts of the preceding text, mention is made of the effect of gas in solidifying metal.   Gases, in particular, hydrogen, nitrogen and oxygen, often readily dissolve in liquid metals.   Nitrogen and oxygen tend to react with a liquid metal to form compounds which may be harmful.   In general, nitrogen causes relatively little trouble; but oxygen can usually form harmful oxides which may render the solidified material unsound either by remaining within it in the form of included particles or grain boundary films or by decomposing during solidification to form bubbles or fissures in the material.

Hydrogen dissolves readily in most liquid metals, separating into the atomic form (H) as it goes into solution. Usually, the higher the temperature of the liquid metal above its melting point the more hydrogen will dissolve, but as the temperature approaches the boiling point of the liquid the solubility for hydrogen progressively decreases. Most industrial melting of metals and alloys has to be done within or near the maximum solubility range; so hydrogen will be absorbed if it is available (e.g., from water vapour or hydrocarbons). Hydrogen is relatively insoluble in the solid material, therefore, as solidification takes place it will evolve more or less rapidly, returning to the molecular form (H_2) as it does so. If hydrogen is entrapped during solidification, the formation of H_2 in the crevices in which it is entrapped can generate quite high pressures and can cause cracking and/or blistering.

Any gas entrapped in solidifying metal can cause unsoundness in the form of microcracks, porosity (numbers of small bubbles), blowholes (very large bubbles or accumulations of gas), cracking, blistering, etc., according to the amount of gas entrapped, the stage at which it is entrapped and its rate of diffusion in the solid. In general, the faster a metal is cooled during solidification the more gas becomes entrapped and the more finely dispersed may be its form.

The effects of absorption and evolution of gas is considered in a little more detail in Sections 3.3.4, 3.4.2 and 3.5.2.

2.5 QUENCH HARDENING AND TEMPERING OF STEEL

As mentioned in Section 2.2.5, steel is basically a binary alloy of iron with a small amount of carbon (up to perhaps 2%) which is usually present in the material as an iron carbide compound, Fe_3C. Generally, steel contains other alloying constituents in addition to the carbon, but, in most steels, the proportions of these are fairly low, totalling perhaps up to about 2%. It contains also small amounts of unavoidable impurities including sulphur and phosphorus. Alloying additions can materially alter the metallurgical behaviour of steel, see Section 2.5.3, but it is easiest to start by considering the behaviour of plain-carbon steels containing the minimum economic amount of impurities and alloying elements.   The nature of iron is dealt with, briefly, in this section, and plain-carbon steel is discussed in a little more detail in Sections 2.5.1 and 2.5.2.   It should be noted that impurities cannot be completely eliminated from iron without raising costs to an uneconomic level.

Iron exists, at room temperature, in a body-centred cubic crystalline form, called α iron and changes at 910°C to a face-centred cubic structure called γ iron.   The only other change within this range is the rather abrupt, almost complete, disappearance of ferromagnetism as temperature rises above 768°C (although the structure remains body-centred cubic, non-magnetic α iron used to be called β iron, but the term has lapsed).   At about 1,400°C, the structure changes back to a body-centred cubic form, called δ iron, before melting at about 1,535°C.   Carbon is almost completely insoluble in α iron and forms iron carbide Fe_3C.   The α iron phase, which can be thought of as pure iron although it does contain a minute trace of dissolved carbon, is known as 'ferrite' and the iron carbide phase as 'cementite'.   The γ form of iron will dissolve carbon, so, on heating into the γ range, the carbide goes into solution in the iron.   This γ phase with or without dissolved carbon is known as 'austenite'.

Rapid cooling of carbon-containing γ iron prevents reformation of Fe_3C and retains the carbon in supersaturated solid solution.   Simultaneously, the structure changes to a somewhat distorted and very stiff body-centred cubic lattice, giving a phase known as 'martensite'. In this state, the structure develops very high strength and hardness (about 850HV with eutectoid composition of 0.8%C) but is very brittle; so, to be made usable, it has to be reheated to allow some of the embrittling distortion to be released by precipitation of iron carbide, i.e., it has to be 'tempered' as described in Section 2.2.5, losing some of its strength and hardness in the process.   Some of the quenching distortion is due to the *expansion* that takes place as α forms from γ iron. This creates a particular problem if adjacent volumes transform at different times and in such conditions cracking may occur.

The amount of carbon present in iron influences the nature of the martensite and the temperatures at which changes take place.   An iron–iron carbide equilibrium diagram, usually called simply 'the iron–carbon diagram', is helpful for understanding, but it must be borne in mind that, although it has its uses, the diagram as such can tell

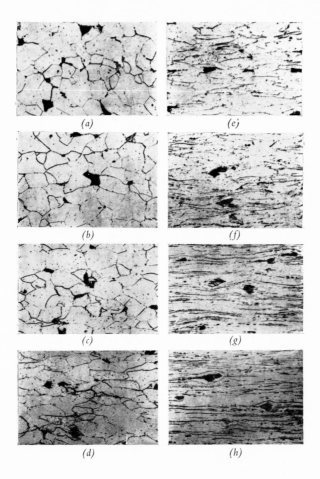

Plate 2.1. Breakdown of the grain structure of a 0·1% C steel with increasing cold deformation. (a) Undeformed. (b) 5%. (c) 10% then in 5% increases up to 35% in (h). (All ×150) (Courtesy: Hanemann and Schrader, 'Atlas Metallographicas')

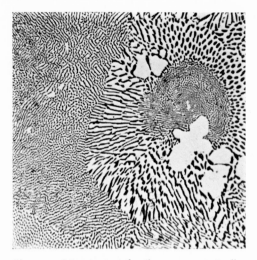

Plate 2.2.   Microstructure of a silver–copper eutectic alloy.
Slight excess of silver (white) giving some primary silver
dendrites (×200)

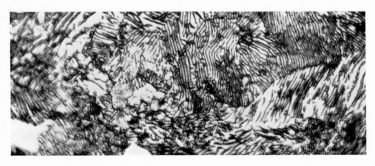

Plate 2.3.   Microstructure of pearlite in a steel of eutectoid composition, showing lamella
of cementite (dark) in ferrite (light) (×500).   (Courtesy: D. L. Thomas)

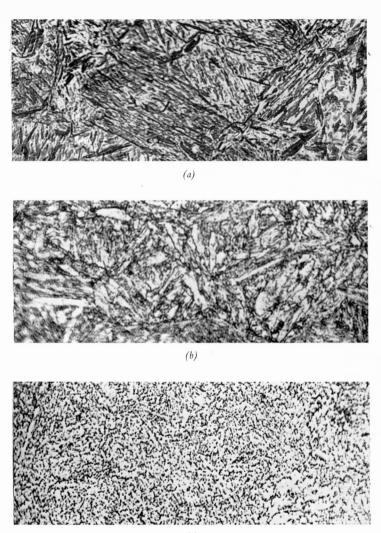

(a)

(b)

(c)

Plate 2.4.   *Effects on microstructure of quench hardening and tempering a plain 0·8% C steel.   (a) Quench hardened showing acicular martensitic structure.   (b) Quench hardened and lightly tempered martensitic structure.   (c) Quench hardened and more deeply tempered giving fine cementite in ferrite (light background), a structure called 'Sorbite'.   (All ×1,000) (Courtesy: D. L. Thomas)*

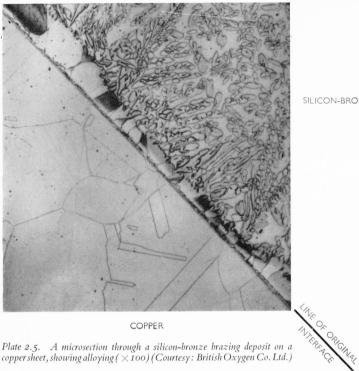

SILICON-BRONZE

COPPER

LINE OF ORIGINAL INTERFACE

Plate 2.5. *A microsection through a silicon-bronze brazing deposit on a copper sheet, showing alloying (× 100) (Courtesy: British Oxygen Co. Ltd.)*

Plate 2.6. *Absorption of the weld face into the metal structure of a resistance butt weld in copper.* *(×40)* *(Courtesy: The Welding Institute)*

nothing about either the quenched, or quenched and tempered conditions of the alloys. Alloying elements added to steel can greatly modify its transformation behaviour and, hence, the form of the diagram.

2.5.1   The Iron–Carbon Diagram

The portion of the iron–carbon diagram relevant to steel is shown in Fig. 2.28. It can be seen that the main feature is a eutectoid reaction at B

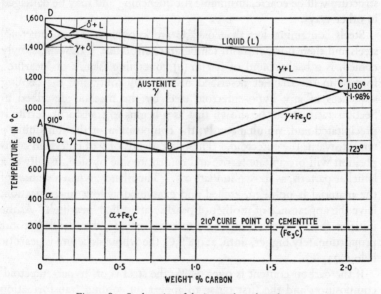

Fig. 2.28.   Steel portion of the iron-carbon phase diagram

(0.8%C and $723°$C).   At this composition, the structure is completely austenitic (γ) above $723°$C and has two phases (ferrite (α) +cementite) below it.   It is apparent that this amount of added carbon has depressed the austenitic temperature of the initially pure iron down to this eutectoid temperature; but the addition of more carbon leads to a rise in the full solubility temperature for carbon until, with 2%C at B, it is up to $1,130°$C.   Carbon tends to suppress the δ change, so, with

0·5%C at D, this change has disappeared. The liquidus is lowered with increasing carbon content and there is a eutectic, with 4·3%C, at 1,134°C, but this is outside the range of Fig. 2.28.

It is possible from the diagram to follow the normal phase changes (the diagram does not represent truly stable conditions) for any one alloy during slow heating and cooling; however, it is probable that the main data to be derived are the temperatures required for normalising, annealing and heating, prior to quench hardening, of particular plain-carbon alloys. All such temperatures should be just a little above the line ABC, but if they are too far above ABC, the structure will be coarse, unsuitable for quenching and may be damaged in other ways.

Steels containing less than 0·8%C are known as 'hypo-eutectoid' steels and their normal room temperature structure, if they are slowly cooled, is a background of ferrite (α) containing islands of 'pearlite', formed in the manner described below. By following the cooling sequences of any hypo-eutectoid steel by the method described in Section 2.4.1, it can be shown that as γ begins to transform, ferrite is precipitated and, simultaneously, the composition of the remaining γ tends towards the eutectoid. These last islands of γ of eutectoid composition will precipitate ferrite and cementite side by side, usually in a lamellar pattern, as shown in Plate 2.3. The islands, when a section of the material is polished, etched and examined at low magnification, have the appearance of mother of pearl so are called 'pearlite'. As the initial carbon content is increased towards 0·8%C, the islands become proportionately bigger, until, at 0·8%C, the whole structure is pearlitic below 723°C.

If the carbon content is over 0·8%, the steel is of 'hyper-eutectoid' composition and the first phase to appear on cooling transformation from austenite will be cementite; but, once more, stable transformation is completed by formation of pearlite. Thus, islands of pearlite are found surrounded by cementite and the nearer the composition is to 0·8% the larger these islands become. Hyper-eutectoid steel is not used so frequently as hypo-eutectoid steel and, even then, the carbon content is rarely more than, perhaps, 1·2%.

Steels containing less than about 0·25%C are called 'low-carbon' or 'mild' steels, those containing about 0·25–0·5%C, are 'medium-carbon' steels and those above 0·6%C are 'high-carbon' steels.

2.5.2 *Martensite, Tempered Martensite and Sorbite—T.T.T. and C.C.T. Diagrams*

Martensite, see Section 2.5, is always likely to form if a ferritic steel (see Section 2.5.3) is heated into the austenitic condition and then cooled rapidly.   Such a formation can develop either as an incidental to some manipulative process, in which case the martensite will be a definite nuisance because of its hardness and brittleness, or it can be deliberately produced in the course of imparting particular properties to the steel when it will usually be an intermediate condition.   A typical martensitic structure of a eutectoid steel is shown in Plate 2.4 (a).

Effective use of the quench-hardening properties of a hardenable steel depends, usually, on the control of formation, and control of decomposition of the martensitic structure.

Control of formation of martensite in a plain-carbon steel is related to two factors: (*a*) rate of cooling, and (*b*) carbon content.

If quenching is not sufficiently rapid, then martensite may either not be formed at all or may form only in small quantities.   Thus, for every value of carbon content there is a limiting rate below which the alloy will not harden effectively.   The interior of a mass of cooling metal cools more slowly than the exterior, the centre cooling slowest and outside skin fastest with uniform conditions of heat conduction from the surfaces.   Therefore, if the centre of a mass of steel is to be quench hardened it must cool faster than the critical rate, a condition which can be achieved only by cooling the outside at a faster rate still.   The thicker the section the faster the outside must cool to give the critical condition at the centre, therefore, the greater will be the contractional stresses between the exterior and the interior and, hence, the greater will be the risk of cracking and the greater will be the residual stresses left after quenching if cracking does not occur.   It can be seen that, with any given quenching method, a limit is set on the depth to which hardening can be achieved in a given mass of steel.   The normal limit is set by the fastest commercially practical mode of quenching, which is quenching into chilled brine.   These limitations on depths of hardening are known as 'mass effect' or 'size effect'.

The lower the carbon content of a steel, the higher is the critical cooling rate; therefore, the lower carbon steels require the fastest cooling.   There is a minimum carbon content (mild steel) below which effective quench hardening cannot be achieved even in the thinnest sections. It should be noted that the term 'effective' has been used several times in the preceding text in relation to hardening by quenching. This distinction is necessary because it is not true that no hardening can occur with cooling rates below the critical one; however, the degree of hardening in this situation is not high enough to justify the common use of this expensive form of heat treatment for this limited result. Even very low-carbon steels are capable of some degree of hardening by quenching, in particular circumstances, giving a situation which can be very embarrassing to some forms of welding.

It is, mainly, because of the relatively high critical cooling rates needed with hardenable plain-carbon steels that they are not in such frequent use for quench hardening applications as are low-alloy steels, specifically developed for the purpose (see Section 2.5.3).

Control of decomposition of martensite is necessary to eliminate both the residual stresses set up by uneven quenching and excessive embrittlement inherent in the martensite type of structure. Both ends are achieved by reheating the quench-hardened material to a suitable temperature, lower than the eutectoid temperature but high enough to allow some internal plastic adjustment that will relieve residual quenching stresses (Section 2.5) and permit the requisite degree of structural adjustment.

This structural adjustment results from the precipitation of cementite in nucleated areas.

In the early stages, the marked acicular appearance of the martensite of a eutectoid steel (see Plate 2.4 (a)) begins to fade, forming a tempered martensite structure in which the cementite particles are just beginning to resolve themselves at magnification $\times 1{,}000$, see Plate 2.4 (b). This structure may have a hardness of about 450HV, a tensile strength of about 100 tonf/in^2, a proof strength of about 80 tonf/in^2 and an elongation of about 5%. In more advanced stages of precipitation, the acicular structure disappears completely as cementite particles become larger and more clearly visible (at $\times 1{,}000$) in a background of ferrite giving a structural state known as 'sorbite', see Plate 2.4 (c). The material in this state will have a hardness of about 350HV, a tensile strength of about 70 tonf/in^2, a proof strength of about 45 tonf/in^2 and an elongation of about 15%. Further tempering would coarsen the structure and bring the properties down to about

300HV hardness, 60 tonf/in² tensile strength, about 40 tonf,in² proof strength and about 18% elongation.

The properties obtainable from a plain-carbon steel depend on both the carbon content and the heat treatment. Quenching at a rate faster than the critical cooling rate gives maximum hardness but lowest ductility and shock resistance; subsequent tempering gives lower hardness but better ductility and shock resistance. Table 2.1 indicates the variation with a typical 0·4%C steel. Maximum hardness is attained in a steel with about 0·9%C. Hypo-eutectoid steels are more ductile than hyper-eutectoid steels.

Table 2.1. VARIATION OF PROPERTIES OF A HEAT TREATED 0·4%C STEEL

Treatment	W.Q.	W.Q. and tempered at			
		200°C	350°C	450°C	600°C
Hardness (HV)	670	620	440	370	270
Tensile strength (tonf/in²)	120	110	92	78	57
Elongation (%)	5	12	14	16	24
Izod impact value (ft lbf)	8	14	10	22	55

W.Q. = water quenched.

Each composition of steel has its own characteristic way of transforming when it is quenched at a given rate from the austenitic state and, at present, the only possible way to summarise the effects of differing conditions is to use one of two graphical methods for recording the results of two types of test, namely; isothermal transformation tests or continuous cooling tests. Curves derived from the former are called 'time-temperature-transformation' (T.T.T.) diagrams and from the latter, 'continuous-cooling-transformation' (C.C.T.) diagrams. Isothermal transformation tests are made on small specimens which have been initially austenitised and then quenched to a predetermined lower temperature and held for differing time intervals before final quenching to room temperature for examination. The times (abscissa) for different degrees of transformation at a given isothermal temperature (ordinate) are recorded. With a plain-carbon steel, a T.T.T. diagram usually takes the form shown in Fig. 2.29 (a) and, consequently, was called an 'S' diagram, but this term has been dropped. Although this type of diagram is useful for comparing steels, it does *not* predict accurately the results of either quench hardening or welding conditions. The C.C.T. diagram, on the other hand, can be used for these purposes. This test utilises differing sizes of round bar to derive cooling conditions equivalent to oil quenching conditions on round

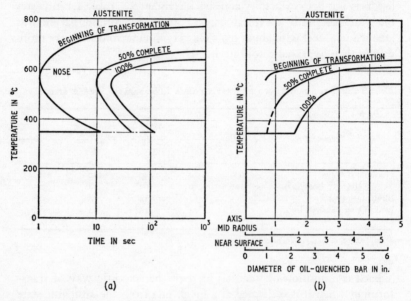

Fig. 2.29. Appearance of T.T.T. and C.C.T. diagram for a 0.35%C carbon steel. (a) Approximate form of T.T.T. diagram. (b) Approximate form of C.C.T. diagram

bar. From these tests, the effects on structure at the middle, surface and half radius, respectively, of the bar are studied and recorded on a graph. Bar diameter is the abscissa and temperature the ordinate as shown in Fig. 2.29 (b). The general form is similar to that of a T.T.T. curve.

2.5.3 *Alloy Additions to Steel*

It is often desirable to adapt the characteristics of a plain-carbon steel to a particular application for which it is not entirely suited. The adaptation might be needed to improve chemical, physical or heat-treatment properties and could be achieved by adding suitable alloying constituents to the basic iron alloy.

A typical chemical requirement is for improvement in the corrosion resistance of steel. This is most often done by adding chromium to a low-carbon steel, possibly in conjunction with some added nickel, to give a 'stainless' steel. Many physical properties may require adaptation. For example, the magnetic properties may be important, so silicon may be added to improve permeability in a transformer steel or chromium may be added, perhaps along with cobalt, to a high-carbon steel to improve its usefulness as a permanent magnet. Resistance to shock loading in the presence of a notch may be important, so manganese might be added. Creep resistance at elevated temperature may be desirable, so chromium, molybdenum, tungsten or vanadium might be added.

Any alloying constituents added to a steel will modify the latter's normal response to temperature change. Changes, such as modification of the eutectoid composition and temperature, raising or lowering of the fully austenitic temperature, raising or lowering of the critical cooling rate, can occur in the presence of one or more appropriate elements. Sometimes the effect is deliberately developed; sometimes it is incidental.

Alloy steels are grouped, roughly, into three classes, 'low-alloy', 'medium-alloy' and 'high-alloy' steels, according to the total amount of alloying addition that has been made. A low-alloy steel generally contains less than about 3% of alloy addition and a high-alloy steel usually has over 10%; but this classification is very vague and gives no guide to the purpose or nature of the alloying. On the other hand, the grouping of steels by purpose, for example, 'air hardening', 'oil hardening', 'notch-ductile', 'corrosion resistant', 'creep resistant', 'high permeability', etc., gives a guide to their application but not too their composition or to their structural characteristics. Generally, a steel is described more fully by using one of the application titles qualified by naming the main alloying additions, and/or, possibly, by naming some structural characteristic (e.g., case hardening 3%Ni steel and Cr–Ni austenitic corrosion resistant steel). However, even this method gives very little guidance on the more detailed qualities of the material and the situation is aggravated by the fact that some materials are well suited to more than one application, each application relying on a different optimum combination of properties. The situation is such that, unless one is very familiar with a particular range of alloys, it is necessary, if accurate metallurgical information is required concerning them, to study in detail composition ranges, T.T.T. and C.C.T. diagrams (if available), mechanical property summaries, etc. This requires a high degree of knowledge and skill, not possessed by the average engineer; therefore, if there is any doubt, a qualified metallurgist should be consulted.

It must be emphasised that no two different alloy steels, even although their compositional differences may seem comparatively slight, will possess a full range of identical properties; there are bound to be some differences in one aspect or another. The whole subject is too complex for detailed discussion here, but it is worthwhile to look at some of the main effects of the more common alloying additions. Aluminium, boron, chromium, cobalt, manganese, molybdenum, nickel, niobium, phosphorus, silicon, sulphur, titanium, tungsten and vanadium, are all commonly found in steels. All the elements mentioned, with the exception of phosphorus and sulphur, are usually deliberate additions.

Phosphorus and sulphur are rarely added deliberately to a steel, but are usually present as impurities. In which case, they should not exceed 0·05% of each, because they could have a marked embrittling effect at all temperatures making the steel dangerous to use, and further, because of scrapping risk, expensive to manufacture.

The useful elements all influence the eutectoid condition of steel, each one lowering the eutectoid composition, but not every one influences the eutectoid temperature in the same way, some, such as chromium, raise it and some, notably manganese and nickel, lower it. The relative effects of different amounts of some additions are illustrated graphically in Fig. 2.30. Simultaneously, these alloys may either retard or accelerate the decomposition of austenite. If the former, then the critical-cooling rate is reduced and heat treatment may be made easier, a situation usually recognisable by increased transformation times and by change in the shape of the transformation diagrams, usually in the form of the appearance of a 'bay' on the 'nose' of the diagram, see Fig. 2.31. If the decompositional change is made easier, the diagram tends to bulge more towards shorter time intervals, indicating that the steel is more difficult to harden; but this

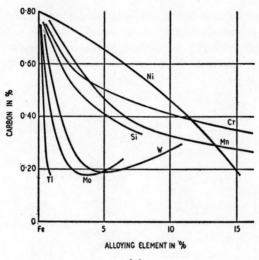

(a)

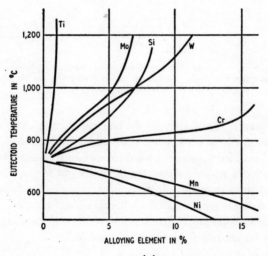

(b)

Fig. 2.30. The effect of alloying elements on the eutectoid relationships of steel (After E. C. Bain, 'Alloying Elements in Steel', p. 127, American Society of Metals (1939)). (a) Effect of alloying elements on eutectoid composition of steel. (b) Effect of alloying elements on eutectoid temperature of steel

effect is unusual among the additions listed above, being caused only by cobalt in certain circumstances.

Some of the elements mentioned above, either alone or in combination in small quantities, have a grain refining effect during casting and

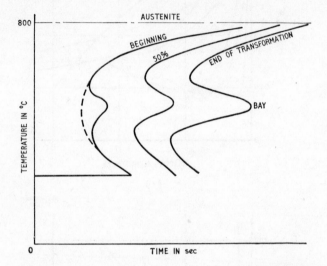

Fig. 2.31.   Influence of alloying elements (e.g., Ni and Cr) in changing the form of the T.T.T. diagram showing the development of a 'bay' on the 'nose' (see Fig. 2.29 (a))

heat treatment.   Thus, they might be utilised primarily for this effect but aluminium is the most usual addition for this purpose.   The latter may also be added to steel intended for surface hardening by nitriding, because it has a strong affinity for nitrogen and forms a hard nitride.

To illustrate the influence of one alloy addition on heat treatment, the quench hardened, and quench hardened and tempered, properties of two steels containing different amounts of nickel and equal amounts of carbon are shown graphically in Fig. 2.10.

2.5.4   *Annealing, Normalising, Stress Relieving and Sub-Critical Annealing*

A few important heat treatments have been mentioned in the preceding sections but they have not been discussed, notably normalising and sub-critical annealing. Annealing is mentioned in Section 2.1.4 but justifies further treatment.

Annealing is, basically, the heating of a metal at a sufficiently elevated temperature for a sufficiently long time to enable the metal structure to attain its optimum stability. It should be noted that perfect stability is likely to be attainable only in a single crystal; and, because of the relatively inferior properties usually found with large grain sizes, it is undesirable in nearly every application. Thus, annealing is usually taken to an optimum degree of stability related to the purpose of the annealing. In the case of a cold-worked metal, the purpose might be recrystallisation to a reasonable grain size; whereas, in the case of a cast ingot, the purpose might be attainment of uniformity of composition and/or structure. The latter two might be achieved by soaking an alloy ingot, or casting, at a suitable solid-solution temperature level for a time long enough to allow diffusion to develop a uniform solid-solution structure before cooling in a manner that will retain a structure as uniform and stress-free as possible. Often, the term annealing is used for any heat treatment process applied to produce softening in a structure.

Normalising is an annealing process particularly applicable to low-carbon and some low-alloy steels. The steel is heated into the austenitic range long enough for the whole structure to austenitise, then the steel is cooled in still air, so that moderately rapid transformation (with its accompanying reformation of grain structure) down to the room-temperature structure gives a relatively fine grain size in the ferrite with a fairly uniform distribution of pearlite. Full annealing of steel by heating within the austenitic range and then slow cooling in the furnace gives a coarser, usually less desirable, pearlitic structure, although it is softer and more stress-free than the normalised structure.

Stress relieving, mentioned in Section 2.1.4, can be achieved most completely by full annealing; but normally, stress-relieving heat treatment is performed at a temperature too low to form austenite or cause much ferrite grain growth. Usually, the purpose of this form of treatment is to allow residual stress to decrease to a tolerable level, without any marked decrease in mechanical strength or change in structure. Such a treatment rarely leaves the structure completely residual-stress-free. It is sometimes called *sub-critical annealing*, or *'process' annealing* if recrystallisation of the α iron is caused.

2.6 METALLURGICAL BONDING IN WELDING

The term 'weld' implies more than a mere sticking together of two pieces of metal, as with a layer of glue. Implicit in the word and clearly stated in most definitions of the technology is the need for a form of true union between the parts that are welded. In metals, such union must be the result of interdiffusion of the atoms of one metal part with those of the other, or atoms of the two parts with the atoms of an intermediate layer of metal differing from both of the main parts. Most welds might be regarded as the latter type, since it is unusual for the material within a weld joint area to have both structure and properties completely identical with those of the parent material.

For interdiffusion to be possible, certain conditions must be fulfilled:

1. Atoms on opposite sides of a metal interface must be intimately in contact with each other so that diffusion distances are small.
2. Diffusion barriers, such as an incompatible continuous layer of oxide, must not be present between the sides.
3. Thermal energy must be applied at an intensity sufficient to promote rapid diffusion.

Welding processes differ widely in the manner in which they fulfil these conditions.

There is another effect, adhesion, which can help to hold surfaces together; but the effect is usually so feeble, compared to true welding, that the conditions in which it operates are generally classed as defective in the welding sense. Adhesion is the sticking together of two intimately touching surfaces by the taking up of a number of the unattached atomic linkages, likely to be available from surface atoms, by pairing of some atoms across the interface. Except in special circumstances, the number of cross linkages is not likely to be very large. This mechanism is probably the main strength factor in a glued joint and in some soldered joints.

In addition to diffusion bonding in a weld, one other strengthening effect is usually present, namely, surface keying. Due to the atomic-scale irregularity of any metal face, it is often easy for mechanical interlocking to take place between two mated surfaces as the asperities of one sink into the recesses of the other. This effect is greatly accelerated in welding processes, even in solid diffusion welding (see below), because asperities on a surface are likely to diffuse most rapidly into a contacting opposed surface. It is impossible to state how much of the strength of a particular welded joint is attributable to keying, but, undoubtedly it does contribute something.

There are three basic types of diffusion mechanism used in welding:

1. Liquid–liquid diffusion.
2. Liquid–solid diffusion.
3. Solid–solid diffusion.

Each of these is discussed below.

2.6.1   *Liquid–Liquid Diffusion: Full Fusion Welding*

One of the most obvious ways to fulfil two of the diffusion conditions mentioned in Section 2.6 is to melt metal into a prepared joint between two pieces of metal. The molten metal, because of the mobility of its atoms, will be able to conform closely to the contours of the joint faces and will also be able to diffuse readily. The situation becomes even more simple if the joint faces themselves are made molten, thus increasing the mobility of their atoms. One problem remains, that of barriers to diffusion; however, if these can be overcome, it is possible to make a 'full fusion' weld or, more simply, a 'fusion weld'; that is, a weld in which 'parent' joint-face metal is melted into, or 'fused' with, the deposited liquid filler metal, or 'weld deposit', that is used to close the gap. The filling metal may be fed in as a separate supply or may be derived, in certain cases, from the design of joint. After union, the joint is allowed to solidify. Two typical joint arrangements are illustrated in Fig. 2.32, one requiring a separate filler metal supply and the other being self-supplying (autogenous).

Solidification is a simple matter in these conditions because nuclei are already present in the unmelted parts of crystals left on the edges of the 'fusion zone', which is the volume of parent metal actually melted into the weld deposit. Columnar crystals readily grow from such nuclei as shown in Fig. 2.33 and the solidified weld resembles a small chill casting bonded into the mould walls, which are thus integrated with the casting. The columnar structure will tend to persist through

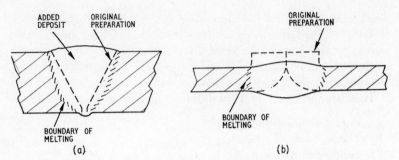

Fig. 2.32. Two types of full-fusion welded joint. (a) Single-V butt, requires extra filler material. (b) Raised edge square butt, this is self-filling

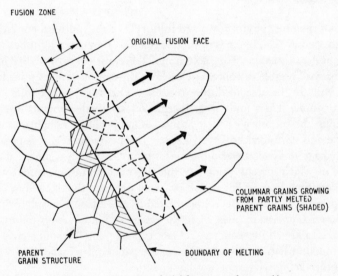

Fig. 2.33. Beginning of solidification in a fusion weld

from one layer of weld deposit into the next layer in a multi-run deposit (e.g., see Vol. 2, Plate 2.1 (b)).

If a material can be cast effectively, it can be fusion welded, provided that proper account is taken of the particular mode of solidification and contraction in welding (see Chapter 3). It is also possible to fusion weld dissimilar metals if they are capable of alloying suitably with each other. Another possibility is the fusion welding of two dissimilar, incompatible, parent materials by using a weld deposit of a third material (e.g., a nickel alloy may be used to bond iron to copper) compatible with both, but it is not always easy to find a suitable third material. The structure of a full fusion weld is similar to that of a chill cast ingot (Fig. 2.5), without chill crystals since effective nuclei are

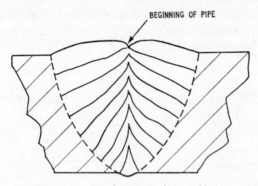

Fig. 2.34. *Pipe formation in fusion welding*

already present. Although chilling is usually very rapid, it is possible, with certain materials having a very steep composition gradient, to develop a central equi-axial crystal zone (Fig. 2.22). In the larger-sized weld joints, it is possible to see the beginnings of a pipe formation (see Fig. 2.24) although this should be avoided as far as possible. In its early stages, the effect shows as a slight notching of the top surface of the weld, as indicated in Fig. 2.34. It should be noted that the meeting of columnar crystals in the centre of a suitably prepared and executed fusion weld does not imply the same degree of weakness as in a casting, since, in welding, there can be more interlocking of the relatively smaller and more twisted crystals. Another helpful factor is that segregation is much less than in a casting, because the grain size is usually so much smaller, the solidification rate is faster and wide compositional differences do not get the same chance to develop.

To be effective, a fusion weld should have the following characteristics:

1. A sound metal structure integrally fused with the parent metal.
2. No sharp corners or crevices.
3. It should be finished with a little 'reinforcement' on the top or outer surface.
4. There should be uniform, full 'penetration' right through the 'root' of the joint, as shown in Fig. 2.35 (a).

If the latter is not achieved (it is often difficult), then the irregular penetration should be subsequently chiselled out to clean metal and covered in with a small 'sealing run', deposited on the underside (see Fig. 2.35 (b)) or alternatively some form of back support facilitating initial full penetration may be used. This difficulty, and certain

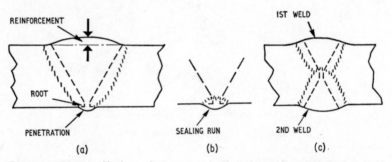

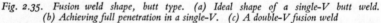

Fig. 2.35. Fusion weld shape, butt type. (a) Ideal shape of a single-V butt weld. (b) Achieving full penetration in a single-V. (c) A double-V fusion weld

problems of distortion, may be overcome by welding just over half way in from each side as shown in Fig. 2.35 (c) giving a 'double' weld.

2.6.2 *Liquid–Solid Diffusion: Brazing and Braze Welding*

Full fusion welding with high-melting-temperature materials involves relatively large amounts of heat input, resulting in serious distortion problems, or excessive welding times, or both. Also there may be undesirable metallurgical effects in certain materials. In suitable circumstances, these problems can, at least, be minimised by a liquid–

solid diffusion system called, according to the type of process, 'brazing' or 'braze' (sometimes 'bronze') welding. These latter terms are misleading as the processes need have no connection with either brass or bronze.

To utilise liquid–solid diffusion, it is necessary to use a filler material of a different composition and lower melting temperature than the parent material or materials, but having a suitable alloying relationship

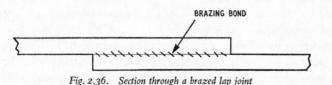

Fig. 2.36. Section through a brazed lap joint

with the latter. The relationship should be such that an alloy with a melting temperature lower than that of the parent material or materials is formed to help the materials bond together. Thus, the parent metal (or metals) temperature is raised just above the melting temperature of the filler metal and the latter is melted on to the joint faces, either directly or by capillary attraction (see below). It is then allowed to alloy with the parent metal by diffusing into it, or vice versa, forming the intermediate alloy or intermetallic compound that creates the bond between the materials when solidification takes place. To be effective the depth of alloy bonding need be no more than a few atomic layers thick with suitable materials. Thus, in looking across the joint faces the sequence of material composition would be: parent material, intermediate alloy layer, filler material, intermediate alloy layer, parent material. The method is often adaptable to joining dissimilar parent materials. A liquid–solid bond face is shown in Plate 2.5 the interdiffusion being clearly visible.

When a lapped joint is used, the lower-melting-temperature filler material is either placed in position before lapping the parent materials (perhaps by 'tinning' the surfaces) or is drawn in by capillary attraction during heating after assembly. The process is then called 'brazing' (its name is derived from the original use of brass for joining mild steel), regardless of the actual materials used. In brazing, a fairly generous overlap has to be made to allow for lack of either penetration of the filler or complete union throughout the joint (see Fig. 2.36).

If the filler metal is deposited in bulk, as in filling the type of joint shown in Fig. 2.32 (a), the process is called 'braze' or 'bronze' welding the latter term deriving from a trade name for the process.

Nucleation for solidification of the filler metal used in brazing or braze welding takes place from exposed faces of those parent metal crystals most suitably orientated for alloying diffusion; but columnar growth is likely to be limited either because insufficient volume of metal is available, as in lap brazing, or because very steep solution gradients develop quickly, thus inhibiting columnar growth and encouraging equi-axial growth.

Soldering in which a very low-melting-temperature alloy (e.g., of tin and lead) is bonded to an unmelted parent material may be regarded as a sub-division of this group; but the type of bond may differ and perhaps may owe more to adhesion and keying than to alloying.

2.6.3   Solid–Solid Diffusion: Solid-Phase Welding

In solid–solid diffusion welding no melting takes place and, for this reason, it is more usually called 'solid-phase' welding.   There are two operations required to complete a solid–solid diffusion process:

1. Mechanical deformation under pressure is used to crush the two weld faces into each other, to give intimate contact.
2. Thermal energy is used to accelerate diffusion so that the structure can recrystallise and grain growth can take place across the interface, which is thus absorbed into the structure. (Plate 2.6.)

Mechanical deformation of weld faces against each other can break down any oxide film that has been allowed to remain on the weld faces and can cause progressively increasing amounts of adhesion to develop, making it possible in certain cases, without further treatment, to attain nearly 100% of the parent metal strength in the joint. However, such strength is achieved only by 80–90% plastic deformation at the interface; an amount impossible to obtain with many metals and, in any case, unacceptable for nearly all purposes, both because of the high stresses required, because of the associated distortion and because of the damage to the parent material structure.   The amount of mechanical deformation caused and the degree of adhesion obtained in the plastic deformation operation has a marked effect on the amount of thermal energy required to induce effective diffusion and recrystallisation across the interface. A suitable balance can give an almost completely invisible weld of 100% of the parent metal strength. It is possible to use a wide range of inversely varying amounts of deformation and heating to achieve very similar strength results. Where thermal energy is readily and economically available, large heat input with minimum deformation can be used and vice versa.

In most solid-phase welding processes, the two operations are performed simultaneously, in comparatively short time cycles, by utilising special machines which perform the required sequence automatically, after being loaded with a set of parent components.

BIBLIOGRAPHY

TWEEDDALE, J. G., *Metallurgical Principles for Engineers,* Iliffe Books Ltd., Chaps. 1–7, 9, 10, (1962).

ROLLASON, E. C., *Metallurgy for Engineers,* Edward Arnold (1948).

REED-HILL, R. E., *Physical Metallurgy Principles,* Van Nostrand, Chaps., 1, 3, 6–14, 16–19 (1964).

CHAPTER 3

Weldability

3.1 SCOPE OF WELDABILITY

As stated in Section 2.6, the term 'weld' implies some form of metallurgical union across the welded joint. In the same section, it is mentioned that welding processes differ widely in the means by which they effect a weld. Thus, the scope of 'weldability' of a material can be considered in two ways: (a) relative to metallurgical aspects, and/or (b) relative to the application of a process.

A perfectly weldable material is one that can be welded readily and effectively by every one of all the possible welding processes; but, industrially, it is more usual to consider weldability of a material relative only to one type of process, or even only to one process, and to neglect all other aspects. Therefore, it is essential that the scope of weldability that is being considered should be clearly defined before ascribing meaning to such terms as 'good weldability', 'poor weldability', 'unweldable', etc. Since weldability is related so closely to the nature of the welding processes and to the amount and nature of weld defects that can be tolerated in particular circumstances, weldability tests are not considered until Vol. 3, Chapters 5 and 6. At this stage, the factors that affect a material's response to the conditions likely to prevail in making a weld will be considered.

Three main problems affect the weldability of a material:

1. Can the desirable metallurgical characteristics of the material be maintained, without undue loss, through to the finished weld?

2. What intensities and types of stress are developed as a result of the welding process during and after making the weld?

3. What is the resistance of the material to such stresses during the time that they operate?

Some of the metallurgical problems, notably of solidification and heat changes are mentioned in Chapter 2 but there are many more aspects to consider. Every method of welding generates its own pattern of stress within the material to which it is being applied and each material is different in its reaction to such stress. The rest of this chapter deals with the more important factors in this very varied field, but it should be noted that space limits consideration to little more than a general survey. Complete coverage needs a comprehensive study of metallurgy, welding plant, welding technique, design and strength of materials, which would require several large volumes.

3.1.1 *Stress in Welding*

Stress in welding can arise in three ways: (*a*) deliberately applied, (*b*) incidentally developed transient, and (*c*) incidentally developed residual stress. The first concerns, mainly, solid-phase welding but the other two may develop in any welding process.

Deliberate stressing is most likely to be compressive with incidentally associated shearing and tension. The latter are likely to be proportionally great if a large amount of plastic deformation is taking place. Plastic deformation is also likely to leave appreciable amounts of residual stress when the applied stress is removed. The intensity of stress will depend on what is being attempted. If the purpose is simply to keep two faces in contact then the stress may be quite low; but, if the purpose is plastic deformation the stress intensity will have to be above the yield stress of the material corresponding to the particular temperature condition during welding. In the latter case, the stress may well start at a high intensity and fall off as temperature rises; but the mode of variation will depend on both the nature of the material and the type of welding process.

Whenever local heating is used in a welding process (and it is local in almost all thermally assisted processes) local stress is generated by differential expansion as heat flows out from the source. Such heat flow is likely to give compression near the source, as local expansion takes place against the resistance of the relatively cold backing material, the stress graduating into tension in the backing material as the latter resists adjacent expansion, as indicated in Fig. 3.1. If the stress is high enough then plastic deformation is likely, first in the compressed

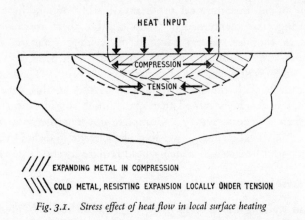

Fig. 3.1.   Stress effect of heat flow in local surface heating

volume, because it is at the higher temperature, subsequently spreading into the stretched volume; but the concentration of deformation will always be greatest in the compression volume.   Thus, when general cooling begins, the compressed volume will be potentially shortened relative to the stretched volume so the stress pattern will progressively reverse as temperature falls towards normal.   At room temperature, a residual tension will be left in the originally compressed area, opposed by a compression stress in the formerly stretched area, as suggested in Fig. 3.2 Any *local* plastic deformation during welding must, inevitably, generate residual stress, the nature and distribution of which will be determined by the mode and pattern of deformation.   A large amount of heat introduced quickly into a small volume will obviously have a different effect from that of a small amount of heat introduced slowly into a large volume.   Thus, if a weld is being made progressively, the speed of progression relative to material quality and thickness will modify the nature and amount of plastic deformation and, hence, of the resultant residual stress.

Another influence, which can affect relative expansion and contraction, is structural change in the material.   Thus, any phase change involving a relatively large volume change in the material is a potential cause of plastic deformation and/or residual stress.

When a fusion weld is being made, there is the problem of the shrinkage of the molten metal as it solidifies against the restraint of the adjacent unmelted metal. This shrinkage can generate relatively high tensile stresses within the weak, cooling, weld deposit and can leave some degree of residual stress.

The more ductile the weld and/or parent material, the more readily can yield take place and, therefore, the lower is likely to be the transient and the residual stresses from any of these causes.

The greatest danger from contraction is most likely to be hot tearing, at temperatures near to the solidus, just when cohesion is being developed between adjacent growing crystals as they grow into

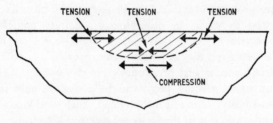

PLASTICALLY COMPRESSED METAL
NOW CONTRACTING AS IT COOLS

Fig. 3.2. *Residual stress resulting from local heating, as in Fig. 3.1,*
until heated metal plastically yields before cooling

contact with each other and, particularly, as they absorb the last traces of intergranular liquid or as the latter finally solidifies into a link between already solid crystals, see Section 2.4.2. Even if melting has not taken place temperatures near to the solidus are dangerous because the grain boundaries are in their weakest state. The greater the range of temperature over which the weakly cohesive state prevails, the greater is the risk of hot tearing. Thus, a material with a wide solidification range (i.e., temperature gap between liquidus and solidus, see Section 2.3) is more liable to hot intergranular cracking than one with a narrower solidification range. Gas evolution from solidifying metal (see Section 3.4.2) at this stage is particularly likely to prevent crystal boundary fusion in some areas and leave microfissures which can act as stress raisers and generate gross macrocracking as temperature falls, or induce failure in service.

3.1.2   Stress Resistance of a Weld Material

Since some degree of heating is used in making most types of welds, the resistance of a material to failure at room temperature is not necessarily a good guide to its weldability with differing processes that may generate differing conditions.   The critical stress conditions for a particular material relative to one process may be set up at temperatures near to the material's melting point; however, the same material with a different welding process may be critically stressed at just above room temperature.   No generalisation may be made except to say that tensile stress is the most likely type of stress to cause weld failure during the welding process and residual tensile stress is the most likely type of residual stress to make an otherwise successfully completed weld unsafe for service.

It can be seen from the preceding statement that a good knowledge of a material's mechanical properties over the whole range of temperature up to its melting point can be a distinct help in assessing overall weldability.   On the other hand, such knowledge is rarely available and, even if it is, the stress conditions likely to be set up by a particular welding process may not be known.   For these latter reasons the stress resistance of a material during welding is very often assessed by some form of empirical weldability test arranged to simulate critical welding conditions (see Vol. 3, Chapter 5).

Failure of a material during welding may take two forms: (a) intergranular cracking and (b) transgranular cracking.   The former tends to happen, mainly, at elevated temperatures when the grain boundaries are likely to be at their weakest, although it can also happen at low temperatures, in a material with a structure in which a brittle grain boundary constituent is present.   Transgranular fracture is nearly always a relatively low temperature phenomenon, because, as material cools down towards room temperature, it tends to become less ductile and transgranular failure more likely.   Usually, it is fairly easy to distinguish a high temperature failure from a low temperature failure, as the fracture facets in the former will show traces of oxidation, by dulling and colour tinting, and the latter will tend to appear clear if not actually bright. Almost invariably it is tensile stress that causes such failure. Both kinds of failure may occur in either the weld deposit (if any) or the parent material near the weld, that is in the heat affected zone (H.A.Z.) of the parent material; but, most commonly, high temperature cracking occurs within the weld deposit. The location of low temperature cracking is dependent on particular circumstances; however, with the fusion welding processes, it tends to be just outside the fusion zone of a fusion weld (e.g., in fusion welding of steel).

A material that cracks readily at elevated temperature is said to be 'hot short' and one that cracks readily at room temperature is said to be 'cold short'. Either weakness can make welding by particular processes difficult and if a material is both hot and cold short (and some are) then it may be almost impossible to weld it effectively.

3.2 INFLUENCE OF TYPICAL WELDING CONDITIONS

Generally, in welding, the metallurgical conditions imposed on a material are likely to be more severe than in more conventional treatments used to fabricate metals into forms suitable for service. When heat is used in welding, the rates of heating and cooling of the material are more rapid than in any other manipulating process. Furthermore, heat is nearly always localised; thus causing severe local distortion and stress problems. Many welded joints are made by progressive welding, which imparts a dynamic effect to the local distortion and stress behaviours and makes the situation still more difficult. If pressure, sufficient to cause deformation, is used in a welding process, the resulting deformation is likely to be much more localised than in, say, a mechanical treatment operation, therefore, this causes its own local distortion and stress problems which have to be solved.

Conditions vary so much between processes that it is impossible to make any reliable generalisations about metallurgical conditions. The best that can be done is to summarise likely effects and later correlate these with particular processes. Basic effects in progressive and solid phase welding processes are reviewed in Sections 3.2.1 and 3.2.2, typical heating and cooling effects in Sections 3.3–3.5, typical structures in Section 3.7 and effects of post-weld heat treatments in Section 3.8.

3.2.1 *Progressive Welding*

Nearly all fusion welding processes, and a few solid-phase welding processes, are progressive in type; that is, the welding begins at one end of a seam or section of joint and is completed by traversing the localised welding conditions along the seam or section. This means that the local heating and cooling, necessary for each relatively small section, is not taking place under constant conditions if rate of heat input, mode of heat input and speed of traversing are kept constant. Such a weld (see Fig. 3.3 (a)) is started on relatively cool material, this necessitates a relatively higher rate of heat input than that required for the rest of the weld in order to attain effective welding conditions in the time available with the chosen speed of travel. Simultaneously,

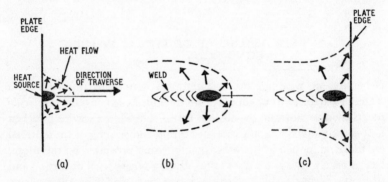

Fig. 3.3.   Heat flow conditions of a progressive fusion weld   (a) Beginning.   (b) Stable
intermediate stage.   (c) End

the rate of cooling will be relatively high. As welding progresses, heat begins to soak into the material adjacent to and ahead of the weld; therefore, the rate of local heating increases and the rate of cooling decreases. After a certain distance of travel, heating and cooling conditions will tend to stabilise relative to the thickness of material, local rate of heat input, etc. (Fig. 3.3 (b)). Stability will be retained as long as the relative rate of spread of heat remains the same, but conditions will change as the end of a seam is approached and heat is no longer conducted away ahead. In this last stage, the weld and the heat affected zones will tend to overheat (Fig. 3.3 (c)). Thus, unless special precautions are taken, the beginning of the seam will be underheated and the end overheated. It is only in certain manual processes that heat input can be readily adjusted locally at the judgement of the operator.

The overall effect of making a progressive weld is a function of mode of heat input, rate of heat input, speed of travel, rate of thermal conductivity away from the heat input zone, thickness of material, area of material around the weld line, degree of mechanical restraint and, sometimes, other factors. High rates of heat input and speed of travel tend to cause fewer metallurgical problems than lower rates of heat input and speed of travel. This is the result of minimising the time available for outward travel of heat, thus narrowing the temperature gradient band and tending to make the conditions more uniform through the thickness. For every progressive welding process, there is an optimum heat input and speed of travel which will give minimum distortion and residual stress; but increase of heat input and speed of travel, beyond the optimum conditions, do not greatly increase distortion and residual stress. More consideration is given to these effects in Vol. 3, Chapter 3.

3.2.2   *Solid-Phase Welding*

Since in most solid-phase welding processes some plastic deformation is necessary for the weld to be effected and the deformation is confined to a relatively small volume of material, it is inevitable that residual stress will be generated. Provided that the material being welded possesses sufficient plasticity to withstand the welding process, it is unlikely that any resultant residual stress will cause either immediate weld fracture or greatly increased risk of service failure. In any event most solid-phase welding processes use heat and are readily adaptable to incorporation of treatment for eliminating residual stress.

3.3   WELD HEATING AND WELD JOINT STRUCTURE

Heat applied in the course of welding has the same effect on metallurgical characteristics of a material as the same amount of heat applied for other purposes, differences arise only from the localised nature of the heating. In the following sections the likely effects of these differences are considered, bearing in mind the kind of temperature gradient (ranging, perhaps, from room temperature up to the melting temperature of the material being welded) that may exist in the vicinity of a weld. The presence of a gradient may mean that a whole range of effects exist alongside each other within a small volume of material; but these effects will divide into groups (a) those deliberately produced within the joint material to effect the weld and (b) those incidentally caused in the parent material within the heat affected zone, as a result of outward flow of heat from the weld. It is most convenient to consider separately weld zone effects only, for each of the three types of weld, and then to consider heat affected zone effects generally, in relation to each of the main phenomena.

3.3.1   *Solution during Heating*

Changing solubility of the constituents within a material to be welded can have a marked effect on the weldability of that material, the actual effect depending on the nature of the change. As indicated above, four aspects will be considered, namely, full fusion welding, brazing, solid-phase welding and heat affected zone behaviour.

Full fusion welding involves the melting of parent material into itself, or into a molten deposit of similar composition, or into a molten deposit of differing composition; however, in each case, it is essential that the main constituents of the material(s) be fully soluble within each other in the liquid state. If no solubility exists then welding is impossible. It is also desirable that the speed at which solution will occur should be high so that the optimum uniformity may be achieved in a short time interval. Should the latter not be possible, then heating would have to be prolonged and many of the economic advantages of welding would be lost. Generally, if liquid solubility is possible, the combined effects of convection and inherent speed of solution diffusion are sufficient to give a high degree of uniformity of composition within the relatively small volumes of liquid metal normally present during a fusion welding operation. This uniformity is attainable with the rapid heating rates normally used in welding mainly because, with nearly every process, there is some inherent stirring action imposed on the metal. The time taken to raise to welding heat a particular segment of material within a fusion zone is never more than a few seconds at the longest.

Brazing can be performed effectively only if there is some degree of solubility between the molten filler material and the heated but unmelted parent material. The degree of solubility need not be great as long as it is sufficient to give an intimate union at the interface and as long as the transition alloy is not excessively brittle or weak. Rapid diffusion is not so essential in brazing because (*a*) the critical layer of metal at the interface is relatively thin and hence diffusion distances are small, usually not more than a few thousandths of an inch at the most, and (*b*) with most brazing processes, time available for diffusion is proportionately great relative to the thickness of the interface. How much interdiffusion takes place whilst the filler metal is molten and how much when it is solidified and cooling is difficult to say, since the proportion will vary with the materials, cooling rate, etc. Solid-state diffusion is likely to be relatively small in its influence, although it may cause a graduation in the structural transition that can help to reduce the effect of any stress concentration and/or cracking that may tend to develop during cooling or in service loading.

Solid-phase welding relies for its effectiveness on diffusion across the interface between the solid materials being welded to each other. These materials must, therefore, be similar in composition or possess some degree of mutual solid-solubility between their main constituents. Melting does not occur in solid-phase welding so liquid solubility is unimportant. Rate of diffusion is not too critical, since, relative to the very short distance over which diffusion needs to take place (a few thousandths of an inch at the most), comparatively long heating times are possible.

Heat affected zone behaviour can be greatly influenced by solid solubility within the parent material. If some or all of the constituents are soluble at elevated temperature in the solid state and almost insoluble at room temperature, then solution will tend to take place within all the material heated above the minimum temperature of solubility. Material at the highest temperature will react most quickly and go farthest in the process, some material will go into complete solution and some will show only shrinkage of the islands of the constituent that tends to dissolve. If precipitation occurs readily on cooling, the effect of the prior solution may be to refine the structure; but, if precipitation is difficult, supersaturation may result, leaving the structure in what may be a very unstable condition. In some cases, a hard brittle zone may be created (e.g., martensite in steel) if care is not exercised (see Sections 2.5 and 3.5).

3.3.2   *Precipitation During Heating*

Generally, precipitation during heating is not a problem in the weld zone of either a fusion or a solid-phase weld, but it can be a problem in the heat affected zone of a supersaturated solid solution material or in partly aged material.   Precipitation of a constituent may lead to some loss of strength and ductility, depending on the material.

If precipitation has harmful effects these can usually be eliminated by post-weld solution treatment and ageing of the weldment, except in the case of unintentionally formed insoluble compounds (see Section 3.3.5.)

3.3.3   *Grain Growth During Heating*

Grain growth during heating is not likely to be a problem in the weld zone of a fusion weld; but it can be a serious problem in the heat affected zone, particularly with materials that do not undergo a grain refining effect during cooling.   With solid-phase welding, grain growth can be a problem in both the weld zone and heat affected zone.

In a material that does tend to undergo grain refining during cooling, the refining effect may be offset by the precipitation of insoluble compounds around the boundaries of the original large grains (see Sections 3.3.4 and 3.3.5).   If, in such a case, the grains do reform on cooling, the boundaries of the initial grains remain outlined in the new structure and their associated weaknesses are likely to be retained.   This kind of weakness often results from overheating a material.

Pure metals that show no allotropic change and alloys that show no change in solubility cannot be grain refined by heat treatment alone (see Section 2.1.4) so the problem of grain growth may be very difficult to overcome.   Mechanical working followed by heat treatment is the usual means for grain refining such difficult materials; but, as a rule, this kind of treatment is not applicable to either the weld zone or the heat affected zones.   The most usual means of approach is to introduce into the composition of the weld material some constituent, perhaps quite minor in quantity, that will inhibit grain growth (see Section 2.2).

When grain growth occurs in a heat affected zone, the largest grains are usually found closest to the weld boundary; but there can be exceptions to this if the material is one that recrystallises at a high temperature and initially has a relatively coarse structure. In the latter case, it is possible to have a band of refined crystals close to the weld boundary bounded on the outside by a zone of larger crystals possibly tailing off in size with distances from the weld boundary. This may happen in the fusion welding of steel when a weld run is

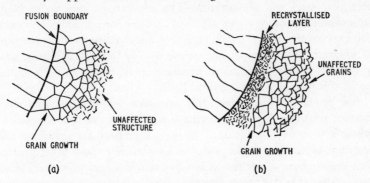

Fig. 3.4.   Grain growth and recrystallisation in the heat affected zone of the parent material. (a) Simple H.A.Z. grain growth.   (b) Partial refining of H.A.Z. grains in an initially course structure

deposited on top of a previous run (e.g., see Vol. 2, Plate 2.1 (a)). The two effects are illustrated in Fig. 3.4.

3.3.4   *Gas Absorption and Oxidation During Heating*

Materials tend to absorb and/or to react with gases in the ambient atmospheres when the material is in either the solid state or the molten state. The higher the temperature is raised the greater are likely to be the rates of absorption and reaction, and these usually show a marked increase as melting occurs.

Oxygen is generally harmful and a metal that oxidises readily is likely to be difficult to weld. If oxidation takes place and does not entirely prevent the formation of an effective weld joint, then harmful oxide will still almost certainly be found either in or near the weld, in the form of oxide particles or as films lying near, or along, the grain boundaries of the material (see Section 3.3.3). Oxygen may also react with particular constituents in certain materials to vaporise the constituents out of the surface and so deplete them, for example, carbon out of steel.

Nitrogen dissolves in many liquid materials and may react with some constituents; but, with most materials, its effect can be disregarded. It does tend to embrittle some materials, notably certain steels.

Hydrogen can be a great nuisance with many molten metals, in particular steel and aluminium. The gas can be picked up from atmospheric steam, or water vapour, or from hydrocarbon compounds that may be present. There is unlikely to be any compound formation of a material with this gas; but, owing mainly to its small atomic size, it diffuses rapidly into many materials, particularly in their liquid states, if it is present in the atomic form.

Full fusion welding. Since this operation involves melting and, usually, relatively high temperatures, it tends to put any material to which it is applied into its most sensitive state both for absorbing and reacting with atmospheric gases. All materials require some protection during heating for fusion welding and many require meticulous protection. Oxidation by direct reaction with atmospheric oxygen is the most obvious form of attack and the rapid formation of an insoluble tenacious oxide film on the fusion face of a joint, whilst the material is still solid, may be sufficient to prevent any union when the metal melts under the skin, since the skin will tend to act as a barrier to the essential interdiffusion. Even if the skin bursts or is deliberately ruptured, the situation may not be greatly improved, fragments of skin being likely to be incorporated in the weld structure as weakening defects. Two possibilities are open: (*a*) complete prevention of oxidation by interposing an artificial barrier such as a film of flux and/or a blanket of inert gas, between atmosphere and material, and/or (*b*) use of a flux to dissolve any oxide as it forms and to make the resultant slag sufficiently fluid and light to float to the surface of the molten metal.

Usually, if a flux can be devised, it is formulated to achieve both possibilities; but, often, an effective flux cannot be devised or it may have unavoidable, undesirable side effects. For example, a flux may be able to dissolve any oxide that forms but yet be unable to make the resultant slag fluid enough to float clear, the slag itself then forming a barrier. Alternatively, a flux may be excessively corrosive in its subsequent action on the cooled material and at the same time be so difficult to remove that its corrosion effects cannot be satisfactorily offset. Fluxes suitable for welding magnesium and its alloys tend to show both these weaknesses. Magnesium is a particularly difficult material in this situation because the material is very reactive with oxygen, the oxide is very persistent, the liquid metal is so low in density that a slag has to be exceptionally light if it is to float clear, and any light active fluxing agents are very corrosive and difficult to remove. An unreactive gas shield is probably the most effective protection and this method is fast superseding the use of fluxes for many forms of fusion welding.

Hydrogen absorption is very rapid with some materials in certain circumstances, more particularly if water vapour is in contact with the surface of the molten metal. In the latter condition, three things tend to happen, the water vapour tends to dissociate, the freed hydrogen tends to dissolve in, and the freed oxygen to react with, the molten metal. However, the real problem of the hydrogen does not manifest itself during the heating process, but during cooling, see Section 3.4.2.

Brazing presents similar if less severe problems, the actual severity depending on the materials being used. In general, with similar materials, the temperature conditions are less severe in brazing, therefore, rather less active slags may be acceptable; but an unreactive gas shield is still the most effective form of protection.

Solid-phase welding, owing to its method of application, causes atmospheric attack to be much less severe than with either of the fusion methods and protection is hardly necessary. Nitrogen is no problem, hydrogen is usually absent, and dissolves relatively slowly even if it is present, while oxidation, although still a problem, is much less critical. The rate of oxidation is likely to be low and, provided existing oxide skins are removed before bringing the weld faces into contact, oxygen is almost completely excluded from between the faces during welding, therefore an oxide-skin barrier is unlikely to reform. If some oxidation does occur, the plastic deformation that is caused in making the weld is likely to disperse the oxide in a relatively innocuous microscopic or sub-microscopic particle form.

Heat affected zone reaction with atmospheric gases varies according to material and process. Fusion processes, which require greater heat input than solid-phase processes, on similar materials, will generate wider heat affected zones and raise the inner parts of these zones to relatively higher temperatures. Nitrogen and hydrogen cause relatively little trouble, even at the higher temperatures, but oxygen may still have its own problems.

Oxygen will usually cause undesirable scaling on the surface of unprotected, heated, solid material alongside a weld joint. Even if serious scaling does not occur, some sensitive materials may be prone to grain boundary oxidation which, although it may not penetrate deeply into the material, may embrittle it by generating sub-microscopic surface fissures that may act as crack initiators. This form of attack tends to be worse if the grain size is relatively large. Oxygen may also react with a surface constituent and vaporise it out of the surface, a reaction well known in steel where, in certain conditions, the skin may be decarburised to an appreciable depth by formation of carbon monoxide gas. The severity of these reactions depends on circumstances, but generally, the protection given to the weld zone will extend far enough to protect most of the heat affected zones. Usually, in conjunction with benefit from protection given primarily to the weld, the time for which the heat affected zone of a properly made weld is maintained at elevated temperature is short enough to keep harmful effects to a minimum.

3.3.5 *Undesirable Structural Reaction During Heating*

Complex alloys are often sensitive to heating, particularly at temperatures near the melting temperature. In these conditions, certain combined constituents may sometimes tend to dissociate and other separate ones may tend to combine to form insoluble compounds, if the material is held too long at the elevated temperature. Steel containing large amounts of dissolved chromium is liable to this kind of reaction, because, in the temperature range 500–800°C, the chromium tends to combine preferentially with the carbon, in the steel, to form relatively insoluble chromium carbides. These chromium carbides tend to form particularly at the grain boundaries and their effect, if the steel is left in this state, is to reduce corrosion resistance, because the adjacent material is depleted of chromium, and to embrittle the material. Fortunately, in this instance the problem may be overcome fairly readily, see Section 3.6 (c). Generally, with the common weldable materials, properly treated and expeditiously welded, structural reactions are not a serious problem.

3.3.6 *Effects of Uneven Heating*

Since many different effects may often be produced in the same material by heating to different temperatures, any unevenness in heating during welding is liable to produce variation between the structures of different welds made as a sequence in the same material or even in the structure of one progressively made weld.   Such variation is always a possible source of weakness particularly in manual welding where the judgement of the operator may vary appreciably even during the making of a single weld.

3.4   WELD COOLING AND WELD STRUCTURE

The cooling soon after its completion of a weld which has been made with the aid of heat induces contractional stresses on the lines suggested in Sections 3.1.1 and 3.2.1, but this cooling may also cause structural changes within the solid material.   Such structural changes can be either harmful or beneficial according to their nature and effects (see Chapter 2).

This section deals with those effects of cooling complementary to the effects of heating considered in Section 3.3.   Thus, grain formation, shrinkage, gas evolution, precipitation, segregation, undesirable reactions and the effects of uneven cooling are each considered relative to solidification and to cooling in the solid, respectively.

3.4.1 *Solidification Shrinkage and Shrinkage During Cooling*

Shrinkage is a problem in both fusion and solid-phase welds but is much more acute in the former, because the amount of shrinkage is likely to be greater and because the prevailing conditions are different.

In fusion welding, the shrinkage begins in the liquid metal as it cools and its effect is influenced by the geometry of the grains that form in the course of solidification.   The grains are usually columnar in form (see Section 2.6.1) but are relatively small in cross-section, compared to similar grains in a casting, furthermore, they nucleate directly from the solid–liquid interfaces of the partly-melted parent metal grains at the boundary of the fusion zone. This latter fact means that: (a) the grain centres are held more rigidly apart from each other, (b) adjacent boundaries generate greater contractional stresses between each other than they would in similar-sized cast structures, because, during cooling, the polycrystalline mass is not able to contract freely in on itself from a free surface; but, (c) solidification is very rapid.

The cross-sectional sizes of the columnar grains are governed partly by the sizes of the parent metal crystals from which they grow and partly by the orientation of these crystals. Not all of the partly-melted crystals will grow, therefore, the cross-sectional areas of the others must expand to fill the space, giving a relatively slightly larger size of grain cross-sections.

Intercrystalline cracking is likely, perhaps in the form of sub-microscopic fissures formed at irregular intervals, in the critically stressed parts of the weld. Such fissures, if they do not open on to an outer surface, can form reservoirs to collect gas that may be evolving from the cooling metal. In a case of this kind, the gas pressure may become a critical factor and cause local extension of fissuring, perhaps to a catastrophic extent, either during cooling, or shortly after the weld is completed, or subsequently, when service loads are applied. In the case where the weld material is inherently brittle, or where a brittle grain boundary constituent is present, shrinkage cracking can be particularly serious in its effects, therefore, special precautions must be taken to avoid its effects. In a particularly bad case of grain-boundary weakness the weld deposit may tend to granulate during rapid cooling. Excessively rapid cooling may tend to induce a generally brittle state in the weld zone of certain materials, making them particularly liable to fracture under the simultaneously-generated shrinkage stresses.

Although, in general, cooling rate does not have a marked effect on the main pattern of solidification in a fusion weld, there can be occasions with some alloy materials, when a slow rate can lead to formation of a central equi-axial crystal zone, such as that shown in Fig. 2.22, but on a much smaller scale. With most fusion welding processes, this effect occurs only in materials with an unusually steep composition gradient. However, there is one effect that may be greatly influenced by the solidification structure and may be significantly affected by cooling rate, namely, the degree of entrapment of non-metallic impurities, such as oxide particles and slag. With many fusion welding processes, these impurities are present in considerably greater quantities than in ordinary casting although, usually, they are in a much more finely dispersed form. Such impurities tend to become entrapped in the solidifying weld metal, particularly at grain boundaries, and add to the weaknesses caused by other effects. Increased speed of solidification may increase the entrapment that occurs. At the same time, it should be noted that the resultant increase in fineness of dispersion of these impurities usually means that their weakening effect on the material is not so great as a similar amount of impurity more coarsely dispersed as in casting.

Heat affected zones of fusion welds can be a serious source of trouble during cooling if the conditions are not carefully controlled. The zone of most rapid cooling occurs just outside the boundary of the fusion zone. It is here that the severest stresses are likely to be set up by shrinkage and at the same time, if the material is one that will quench harden (e. g., heat treatable steel), this is the zone in which hard, brittle phases are most likely to form, see Plate 3.1. The combination of these two effects is likely to make welding difficult in a susceptible material. If grain growth is excessive during heating, grain boundary fracture during cooling is always a possibility, particularly if the grain growth is associated with grain boundary oxidation or precipitation of a brittle constituent.

Brazing and braze welding are both liable to contraction problems similar to those in fusion welding, but to a much lower level of severity, even with the comparatively large volumes of solidifying material used in braze welding.

Contraction in solid-phase welding is not usually so serious as in other forms of welding. For one thing, the solidification contraction is absent so the total range of contraction is not nearly so great with similar materials. Another factor is that external pressure is usually applied during contraction, thus helping both the contraction process and the compacting of the structure. Normally, too, there is very little gas evolution during solid-phase welding. The heat affected zone of a solid-phase weld is not likely to give much trouble unless excessive grain growth or undesirable structural change has occurred during heating; nevertheless, neither of these is as likely to occur as in the heat affected zone of a fusion weld in the same material.

3.4.2 *Gas Evolution During Cooling*

In general, gas evolution is not a serious problem in either a solid-phase weld or within the heat affected zone of a fusion weld; but it can be very serious in the weld zone of fusion and braze welds and occasionally also in the heat affected zone of a fusion weld.

The gas most likely to cause trouble is hydrogen because this gas, during heating, may go into copious solution in many molten metals (see Section 3.3.4) and then be rejected during cooling. If rejection takes place rapidly at the solidification face and the remaining molten metal is still very fluid, the gas will bubble through the metal and escape to the atmosphere. In this situation, unless the gas is mechanically trapped for some reason, it causes little trouble and may even assist in stirring the molten metal and breaking down compositional gradients (see Section 2.4.1) at a time when no other stirring action, apart from convection, may be operating. However, if evolution continues in the solid metal behind the advancing solidification face, or if the molten metal is viscous, gas may become entrapped between crystals (see Section 3.1.1) and in the viscous molten metal. In either case, it may remain as microcracking, usually between grains, invisible to the unaided eye when the macrostructure is examined; or it could occur as porosity (i.e., as groups of tiny bubble holes) within the solidified metal; or it may gather into larger less regular cavities called 'blowholes' (usually in welding a cavity with a diameter over $\frac{1}{16}$ in is classed as a blowhole) or as a 'wormhole' which is sometimes called a 'pipe'. This last term although very descriptive must not be confused with casting 'pipe' so it is safer to refer to this weld defect as a 'wormhole'. A wormhole is a tubular cavity, usually not very long, with rounded ends, formed by a bubble freezing at one side into the solidification face and then, as gas accumulates, expanding on its opposite side into the molten metal keeping pace with the advancing solidification as shown in Fig. 3.5 and Plate 3.2. In certain circumstances, wormholes can grow to quite considerable lengths.

It is possible for hydrogen to be trapped in the weld metal of a fusion weld and, subsequently, for some to diffuse into the heat affected zone, where it may develop microcracking.

As hydrogen diffuses out of a material into a cavity, it combines into the molecular form and in doing so can generate very high pressure.

This is particularly dangerous if it occurs entirely in solidified material.

Other gases do not cause so much trouble as hydrogen, either because they do not go into solution or because they tend to form harmless permanent compounds that are retained within the weld metal. If a harmful compound tends to form (as with oxygen) either it must be prevented from forming or be fluxed out if it does form so that it is not entrapped in the solidifying material. Nitrogen is a relatively insoluble gas and, in casting, can be deliberately bubbled through molten metal (e.g., aluminium) to collect and remove dissolved hydrogen; but such a treatment as the latter is not normally practicable

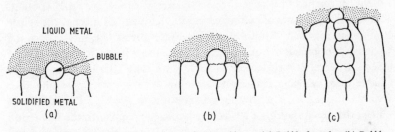

Fig. 3.5. Mode of formation of wormhole in fusion welding. (a) Bubble formed. (b) Bubble partly trapped. (c) Extended bubble completely trapped to form wormhole

in welding. Nitrogen can be a nuisance in the fusion welding of reactive metals and, occasionally, when fusion welding steel, if normal simple precautions are not observed.

3.4.3 Structural Changes During Cooling

As suggested in Section 3.2, the cooling rates encountered with most welding processes tend to be relatively high and, in many cases, if precautions are not taken to reduce them or to offset their effects, harm may result.

According to the type of welding process and the material being welded, the cooling process during either a fusion- or brazing-type weld may influence, (a) the solidification process (see Section 2.4) and hence the resultant macrostructure and microstructure within the weld zone, and (b) the microstructure and possibly the macrostructure within the heat affected zone. With solid-phase welding, the cooling may effect, (a) the microstructure of the weld zone, and (b) the microstructure of the heat affected zone.

In all the processes, the effects of cooling on unmelted but heated metal and metal that has just solidified are similar to those cooling effects described in Sections 2.2.1–2.2.5 for similar initial conditions of the hot material. Thus, solid solution supersaturation, quench hardening and partial suppression of precipitation can all occur during weld cooling in the appropriate conditions.

Of these effects only quench hardening is likely to give much difficulty, but it causes some of the most serious problems in the fusion welding of steels, see Sections 3.5 and 3.6. Probably the formation of a brittle structure in the weld deposit and/or the heat affected zone of a weld, in conjunction with the severe stresses caused by rapid cooling, are the major causes of cold cracking in welds and of subsequent brittle failure in service. Another possible cause with some materials is the partial suppression of precipitation, making overageing (see Section 2.2.2) embrittlement possible with time after cooling. In the latter case, cracking is likely to develop some considerable time after weld completion or when the weld is put into service. Particularly in steel, the precipitation of molecular hydrogen leading to the formation of microcracks is a cause of brittle failure (see Sections 3.4.2 and 3.5.2) of this kind.

If any of these effects are likely to occur, then it is essential that precautions should be taken either to prevent them during the making of the weld or to eliminate their harmful effects after welding. The former is most readily achieved by reducing the cooling rate below the critical value, perhaps by preheating the whole mass of weldment or by other adjustment to heat flow effects in the weld (e.g., a higher heat input usually gives slower cooling). Another precaution may take the form of a stress relieving and/or structural adjustment by heat treatment applied soon after the weldment is complete (see Sections 2.1.4 and 2.5.4).

3.4.4   *Cooling and its Effects on Incidental Reactions*

Sometimes, in making a weld, notably a fusion weld, due to breakdown of protection or other causes, undesirable compounds (e.g., oxides) may form or undesirable reactions may take place (e.g., formation of chromium carbide in alloy steel) (see Sections 3.3.4 and 3.3.5). If such a reaction is irreversible, the cooling process is likely to retain all of the undesirable product in the weld and cause local loss of properties. Even although a reaction may be reversible on cooling, the rapid rate of cooling is likely to inhibit the reversion and again result in retention of the undesirable phase, with similar harmful effects. Subsequent rectifying treatment is sometimes possible; but the only really safe procedure is to prevent the undesirable reaction at the stage when it is likely to take place.

3.4.5 *Effects of Variation in Cooling Rate*

If the speed of cooling is likely to influence the structure of a material, either in a weld zone or the heat affected zone of a weld, then variation in cooling rate is likely to cause variation, usually undesirable, in the material structure. When the weld is made automatically in one simultaneous operation, the cooling conditions are likely to be fairly consistently controllable; but, where the weld is made progressively, the chances of variation increase considerably and are at their worst in a manually-made progressive weld which is carried out on site in the open air. Variation in atmospheric temperature, wind velocity, etc., can all influence cooling rate and therefore affect the structure of a sensitive weld material. Even indoors draughts can affect cooling rate. Furthermore, any progressive weld will have differing cooling rates acting at the start and finish of each weld run because of the differing initial and final heat flow conditions (see Section 3.2.1).

With some processes and materials, it is possible for the cooling rate to influence grain size, particularly in the heat affected zone, since a slower cooling rate means a longer time for affecting material within the elevated temperature range and also a widening of the heat affected zones to affect more material.

The undesirable effects of variation in cooling rate can sometimes be eliminated by a post-weld heat treatment, such as annealing; but, if cracking has occurred, say as a result of the formation of local hard zones, any cracks must first be cut out and repaired.

3.5 COOLING EFFECTS IN WELDED PLAIN-CARBON AND LOW-ALLOY STEEL

In Section 2.5, the effect of quench hardening steels from temperatures above their critical temperatures is considered. Any cooling at a rate faster than the critical rate that takes place within the vicinity of a weld in a quench hardenable steel, where the initial temperature is above the critical temperature and the final temperature significantly below it, will result in the formation of martensite. This martensite is brittle, hard and likely to be subjected to considerable stress which may be sharply accentuated by a delayed γ–α change in a closely adjacent zone (the γ–α change involves an *expansion* of the lattice) as a result of temperature gradient. Also the lower the transformation temperature, the greater is the effect of the volume change on the more rigid lattice structure during the change, so the greater is the tendency to crack.

Hence, the easier it is to quench harden a particular type of steel the more prone that steel is to weld cracking so the greater the care required in welding it. Usually, solid-phase welding conditions are automatically controlled and can be readily adapted to avoid the formation of martensite; therefore, with forethought, brittle zone formation is not a serious danger when using this type of process on a hardenable steel. Fusion and braze welding are not so readily adaptable to control in this respect and martensite formation can be a serious problem. The problem is most serious when a run or spot of weld metal is deposited on relatively thick cool plate with a relatively large thermal capacity, as with multi-run type fusion welds where a thick section is built up by a series of superimposed weld deposits of small size relative to the thickness. The situation becomes progressively worse with decrease in the relative size of the weld.

Since the danger of martensite formation increases with hardenability and with the brittleness of the martensite formed, so the difficulty of welding increases with the carbon content of plain-carbon steels and with the equivalent hardenability of hardenable alloy steels. Because the hardenability of low-alloy steels varies with alloy content, a simple approximate relative measure has been devised in terms of the 'carbon equivalent' (C.E. or C_E). This equivalent is the sum of the equivalent effects on hardenability of each constituent compared with carbon, some constituents having a low equivalent (lower hardenability) and others a high equivalent (greater hardenability). By applying the appropriate correction factors, shown below, the plain-carbon steel composition roughly equivalent in hardenability to a given low-alloy steel may be found and this can serve as a first guide to weldability of the alloy steel.

$$C_E = \%C + \left(\frac{\%Mn}{20} + \frac{\%Ni}{15} + \frac{\%Cr + \%Mo + \%V}{10} \right)$$

Other formulae take account of elements such as copper and phosphorus but usually the quantities of copper and phosphorus in a steel will be less than 0·5% each, so their influence can safely be ignored in an approximation such as this. It should be noted that the C_E is a rough guide only and that it becomes increasingly unreliable as alloy content increases. Other methods of indication have to be used with high-alloy steels (see Section 3.6.1).

As a rule, when fusion welding, it is unsafe to weld thick sections of plain-carbon steel containing much more than 0·25%C_E in cold weather out of doors (this usually means arc welding, q.v.) without the precaution of preheating to retard the cooling rate. As a rough guide, Table 3.1 gives some approximate preheating temperatures for severe welding conditions such as a small weld being laid on a large thickness of material.

Table 3.1. APPROXIMATE PREHEAT TEMPERATURE FOR PLAIN-CARBON AND LOW-ALLOY STEELS

C_E%	Temperature °C
0·40	50
0·45	100
0·50	150
0·55	200
0·60	250
0·65	300

It must be understood that although a steel is classed as 'mild' (see Section 2.5.2) it does not mean that no quench hardening can take place, but only that normal quenching is unlikely to be severe enough to give a brittle, martensitic structure. It is possible with some 'mild' steels in certain conditions to get a critical degree of hardening in the heat affected zone close to the fusion zone.

For a more precise understanding of the weldability of an alloy steel, a detailed knowledge of both the relevant welding conditions, and the T.T.T. or, preferably, C.C.T. diagrams of the alloy is essential.

3.5.1   *C.C.T. and T.T.T. Diagrams and the Weldability of Steel*

Probably, the best guide to the weldability of a particular alloy steel is
its C.C.T. diagram (see Fig. 2.29 (b)) since this diagram is a record
of the transformation behaviour of that steel under continuous cooling
conditions that can be correlated fairly closely with the kind of
continuous cooling occurring in the vicinity of a weld.   From such
a diagram, it is possible to determine whether or not martensite
or a brittle structure is likely to form under given welding conditions.
The further to the right and the lower the curves on the diagram the
more hardenable the steel and the more difficult the welding.

If a C.C.T. diagram is not available a T.T.T. diagram (see Figs.
2.29 (a) and 2.30 (a)) will give a good guide, if not so accurate as the
C.C.T. diagram.   In this case, the welding becomes more difficult
with the lowering of any 'nose' on the diagram (see Fig. 2.31) and the
increase in the time span before transformation begins, since these
features indicate that the transformation is likely to be delayed, under
continuous cooling conditions, to a low temperature at which the
metal will be relatively brittle.   Most severe weld cracking tends to
occur when transformation takes place below 300°C.

Although basic weldability for given welding situations may not be
assessed directly from either type of diagram, without knowledge of
the equivalent weld cooling conditions, it is possible to compare the
relative weldability of two different alloys for similar welding purposes by comparing the relative positions and sizes of their respective
curves.

3.5.2   *Hydrogen in Welded Plain-Carbon and Low-Alloy Steels*

Atomic hydrogen (H) is relatively insoluble in α iron, moderately
soluble in γ iron and very soluble in liquid iron.   If hydrogen is
dissolved either in γ or liquid steel during the heating stage in making a
weld, then the normal weld cooling process is certain to entrap a
considerable proportion of the hydrogen in a state of supersaturation.
The situation is not usually difficult in solid-phase welding because
hydrogen is not likely to be present near the weld faces; but, in some
fusion processes, hydrogen can be a serious problem, particularly if it
is present in the atomic form near the liquid surface, say as a result of
the cracking of water vapour or some other hydrogen-containing

compound.   After cooling, any excess hydrogen atoms present in the metal are likely to try to diffuse to a free surface of the metal where they can combine into the molecular form (H_2).   Unfortunately, some of these free surfaces may be the walls of totally enclosed intergranular fissures, porosity cells or blowholes.   The formation of molecular hydrogen within such enclosed spaces can lead to the generation of very high pressures, which can cause a crack to spread or may start a crack in the wall of the cavity.   Molecular hydrogen does not readily diffuse through solid iron so a generated pressure of this kind is retained in the structure and can help initiate a fracture in service if it does not start one immediately.   Molecular hydrogen formed in this way in microfissures or small pores is the likely cause of the 'fisheye' form of localised brittle fracture often seen in the fracture faces of certain weld-deposited steels (see Plate 3.3), although other forms of microfissuring may also cause this kind of fracture in a weld deposit.

Supersaturation with atomic hydrogen can be a cause of brittle fracture of steel at, or slightly above, room temperature.   This brittle fracture does not occur rapidly under impact, as with notch-sensitive steels, or necessarily at lower temperatures, but takes place under steady working loads.   It is believed to be caused by diffusion of hydrogen into strained metal at the sharp tip of a stressed fissure or crack, a critical concentration of hydrogen having to build up before the crack can spread for a short distance in a brittle manner out of the hydrogen concentration.   Continued diffusion leads to more cracking and so on to total failure.   Naturally, this condition is dangerous and must be avoided, either by preventing absorption of hydrogen in the first place, or by aiding the hydrogen to diffuse out of the metal.   The latter end can be attained with most steels (e.g., mild steels) by allowing time for a sufficient amount of hydrogen to diffuse out before putting the weld into service.   Most authorities recommend a minimum of three days for this, but some say 24 hours is sufficient.

Higher plain-carbon and alloyed steels can be very sensitive to hydrogen and may need post-heating after fusion welding, to perhaps 400–500°C, to aid diffusion; but, it must be kept in mind that elevated temperature may in some cases reduce the ductility of the material and increase the danger of cracking under residual stresses.   Another danger with these steels can be diffusion of hydrogen from the supersaturated weld deposit into the heat affected zone where the pressure built up by the hydrogen in the vicinity of a hard zone is likely to accentuate 'underbead cracking' (i.e., cracking in the heat affected zone ust outside the fusion zone of a fusion weld, see Plate 3.1).

3.6 WELDABILITY OF HIGHLY ALLOYED STEELS

In the preceding section, the inverse correlation between the weldability and the hardenability of plain-carbon and low-alloy steels is considered and the conclusion drawn that, for these steels, the carbon equivalent is probably the best guide to weldability. With more highly alloyed steels, hardenability may be negligible (e.g., austenitic steels) or less useful as a guide, so other factors have to be considered. Highly alloyed steels are utilised mainly for purposes other than simple mechanical strength, the most obvious possibilities being corrosion resistance, heat resistance and abrasion resistance. Steels for such applications tend to group themselves into three main metallurgical groups; (a) martensitic steels, (b) ferritic steels and (c) austenitic steels, according to the state in which the steel tends to exist at room temperature after relatively slow cooling.

(a) *Martensitic steels* are likely to contain a high proportion, up to 16%, of chromium and, with the larger quantities, are completely martensitic even after very slow cooling. Depending also on the carbon content, increasing quantities of chromium make martensite formation increasingly easy and the need for precautions when welding progressively greater. Chromium is very prone to combine with carbon and/or oxygen so these two elements must be strictly controlled during fusion welding.

(b) *Ferritic steels* usually contain over 16% chromium and cannot be quench hardened, because the $\alpha \rightarrow \gamma$ change is completely suppressed, so that the steel remains body-centred cubic right up to melting temperature. These steels may fusion weld relatively easily but, depending on the other alloys present, they may be prone to the formation of a hard brittle phase, called 'sigma' phase, when slowly cooled or heated into the range 400–550°C for prolonged periods of time. Sigma phase tends to give cold cracking in the weld deposit and heat affected zones. Grain growth is a problem with completely ferritic steels at temperatures above 1150°C.

(c) *Austenitic steels* usually contain either high proportions of chromium and nickel, or nickel alone, in excess of 25%, or may be alloyed mainly with manganese. The first type is intended primarily for corrosion resisting applications and the last type for abrasion resistance. The middle type has various limited uses. Several other alloying elements can be used to influence the properties, the essential feature being that the $\gamma \rightarrow \alpha$ change is completely suppressed by air cooling so that at room temperature the material, slowly cooled or quenched, remains face-centred cubic in structure. All have a tendency to high temperature hot shortness.

Of the chromium–nickel steels, probably the best known is the 18%Cr–8%Ni alloy series. These fusion weld fairly easily except for the tendency to 'weld decay'. Weld decay is the tendency to form chromium carbide around the grain boundaries, mainly in that part of the heat affected zone which is raised into the temperature range 500–800°C, causing greatly reduced corrosion resistance (see Section 3.3.5). This weld decay can be eliminated by subsequent heating to 900°C or it can be prevented by adding to the alloy appropriate amounts of niobium or titanium to form stable carbides that will prevent the chromium picking up carbon. Niobium is the more useful additive in that it is more easily incorporated in a fusion weld deposit, but it may increase the tendency to sigma phase formation. The risk of sigma phase being formed is present with all these alloys if heating in the range 400–500°C is prolonged.

The manganese-containing alloys usually contain about 1%C and 13%Mn and are austenitic when cooled from the liquid at the rate normal in the fusion zone of a weld. However, reheating, such as always happens somewhere in the heat affected zone of a fusion weld, into the range 425–750°C can lead to the formation of martensite; thus, a layer of martensite, with its attendant cracking tendency, is likely to be produced in the heat affected zone. The material also tends to transform if it is plastically deformed but this is unlikely to affect the fusion weldability to a marked extent. Nickel is sometimes added to this type of steel to give more stability to the austenite.

Weldability of alloys in the three groups is not easy to assess since there are nearly always more elements present than those from which the basic character of the alloy is derived. Carbon equivalent is almost meaningless in this context, so other means have to be found to give a guide to weldability. Such a guide is particularly important where a highly alloyed steel of one composition is being fusion welded with an alloy of markedly differing composition, as may happen, for example, if hydrogen pick-up is a problem as in welding say a martensitic steel. In this case, the use of an austenitic steel weld deposit may be an advantage since (a) the austenitic structure is ductile and therefore has a greater tolerance for hydrogen and (b) it can retain hydrogen more readily than martensite or ferrite and, therefore, minimises the diffusion of hydrogen outwards into the heat affected zone.

One guide has been found, primarily for corrosion resistant steels, that uses the concept of nickel and chromium equivalents. This guide is considered in the next section.

3.6.1 *Nickel and Chromium Equivalents—The Schaeffler Diagram*

Consideration of the preceding section will indicate that (a) nickel as an alloy addition to steel tends mainly to promote the formation of austenite, (b) chromium tends mainly to promote the formation of ferrite, and (c) intermediate smaller combinations of the two promote martensite formation. Furthermore, in Sections 3.4.3, 3.5, 3.5.1 and 3.6, it is shown that austenite is particularly prone to hot intergranular cracking during fusion welding, martensite is particularly prone to cold transgranular cracking and ferrite is particularly liable to encourage sigma formation, with its attendant risk of intergranular fissuring, these effects varying fairly predictably with the corresponding proportions of nickel and chromium. Thus, if the total effect of the constituents in a particular steel can be found in terms of nickel and chromium equivalents, it becomes possible to make a fair prediction of the metallurgical characteristics of the alloy and its probable weldability. It also becomes possible to predict the likely behaviour of intermediate alloys formed between a parent metal of one composition and a weld deposit of different composition.

A. L. Schaeffler developed from some work by E. Maurer a system of nickel and chromium equivalents and recorded their effects on an approximate graphical basis in the form of the *Schaeffler* diagram (see Fig. 3.6). In Table 3.2, the conversion factors for different alloying elements are given and it is a simple matter, knowing the composition

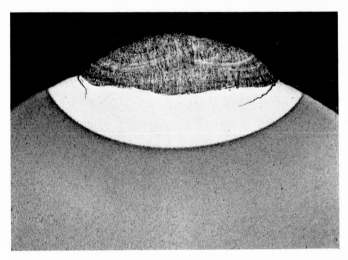

Plate 3.1. Martensite formation (white area) under an arc weld bead laid on a cold alloy steel bar. Transverse section (×3). Note cracking

Plate 3.2. A view of a small wormhole in a fracture face of a broken mild steel tensile test specimen (×20)

Plate 3.3. A pair of 'fisheyes' in a fracture face of a broken mild steel tensile test piece (×20)

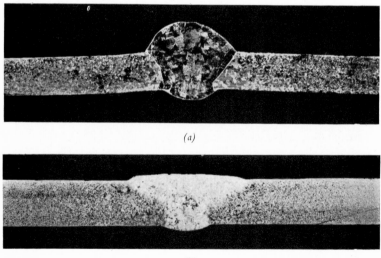

(a)

(b)

Plate 3.4. Effect on structure of oxy-acetylene fusion welding (slow rate of heat input). (a) Oxy-acetylene welded single-V butt weld in copper. Note grain growth in heat affected zone and large grains in weld deposit (×3½). (b) Oxy-acetylene single-V butt weld in mild steel sheet. Note the grain coarsening adjacent to the weld and the spread of the heat affected zone (slightly darker area outside the weld deposit) (×3)

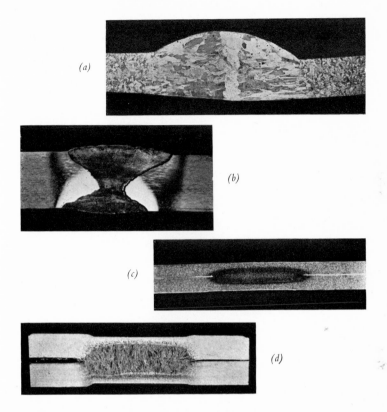

Plate 3.5. Effect on structure of high rate of heat input in fusion welds. (a) Arc welded copper (×5). (b) Mild steel arc welded butt joint (two passes) in plain medium-carbon steel. Top run made first, note the tempering of martensite in its heat affected zone. Bottom weld run second, note untempered martensite (white) in heat affected zone. Cracking started in the initially untempered martensite of the first heat affected zone but other incipient cracks can be seen (×2). (c) Resistance spot welded aluminium (×2½). (d) Resistance spot welded low-carbon steel (×1½). (Courtesy: The Welding Insitute)

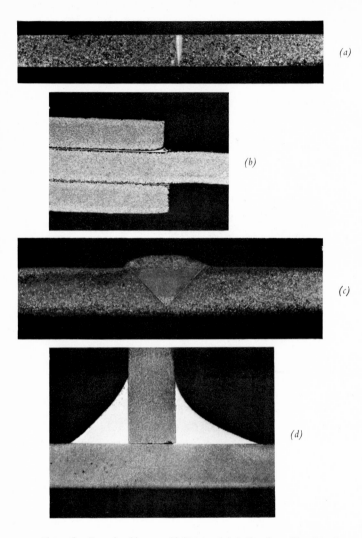

Plate 3.6. Brazed and braze welded joints. (a) A silver brazed butt joint in copper sheet. Note that use of this design of joint is not generally recommended (× 1½). (b) Copper brazed joint in mild steel (× 4) (Courtesy: The Welding Institute) (c) Oxy-acetylene braze welded single-V butt weld with brass in copper sheet (× 2). (d) Oxy-acetylene braze welded T-joint in mild steel sheet. Note the slight mis-shaping of the right hand fillet and the beginning of capillary attraction under the slightly opened end of the vertical sheet. Note also the absence of thermal disturbance to the structure of the steel (× 3)

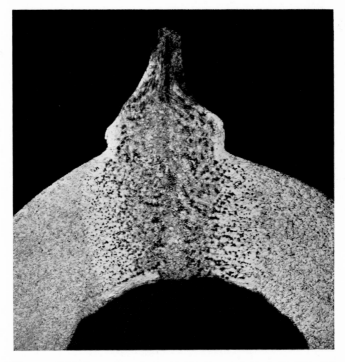

Plate 3.7. Solid-phase resistance butt welded joint in a steel ring (×6). Note that the darker area near the joint face is due mainly to disturbance to the fibre structure of the material. Compare with Plate 2.6, a similar joint in copper. (Courtesy: The Welding Institute)

(a)

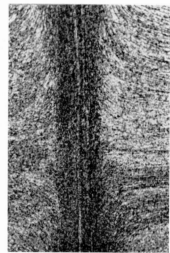

(b)

Plate 3.8.   Cold pressure welded joint in aluminium.
(a) Macrosection of weld (×3).   (b) Microsection at weld face
(×50).   (Courtesy: The Welding Institute)

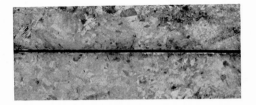

(a)

(b)

Plate 3.9. Soldered joints. (a) A soldered lap joint between copper sheet with lead–tin solder. Note uneven grain structure of the copper owing to uneven control of the heat source causing local overheating (×3). (b) The interface of a soldered joint between a copper–beryllium alloy bottom and 60% Sn 40% Pb solder. Note the width of the copper –tin intermetallic layer (white) which is unusually excessively wide owing to either overheating or prolonged heating. Normally the intermetallic layer is so thin that it cannot readily be resolved (×440)
(Courtesy: Fry's Metals Ltd.)

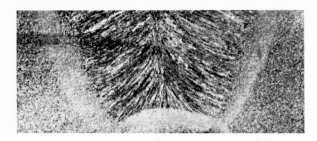

(a)

(b)

Plate 3.10. Structural stabilisation of a single-V fusion weld in low-carbon steel plate. Weld machined flush with plate faces ($\times \frac{3}{4}$). (a) As-welded, note columnar structure. (b) Normalised, note traces of columnar structure still remaining

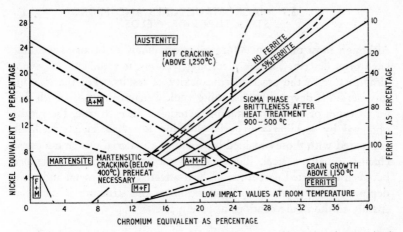

Fig. 3.6. Schaeffler diagram for alloy steel welds, showing possible welding dangers related to composition

of a particular steel, to convert that composition into an equivalent nickel–chromium composition, which can then be located on the diagram and will indicate the probable behaviour of that steel under welding conditions. By locating two dissimilar steels on the diagram and joining their two points, the joining line will give a guide to the likely variation in compositional effects across a fusion transition between one, if it is a parent metal, and the other if it is a weld deposit.

It must be emphasised here that it is not claimed that this method does any more than give a rough guide which can serve as a useful starting point for a weldability study. Within those limits it is a very useful aid.

Table 3.2. NICKEL AND CHROMIUM EQUIVALENTS OF ALLOY ADDITIVES IN STEEL

Additive	Ni Equivalent	Cr Equivalent
Carbon	× 30	—
Chromium	—	× 1
Manganese	× 0·5	—
Molybdenum	—	× 1
Nickel	× 1	—
Niobium	—	× 0·5
Silicon	—	× 1·5

3.7   MACROSTRUCTURES AND MICRO-
STRUCTURES OF WELDS

Although most of the metallurgical effects likely to occur in welding
have been described in the preceding sections, it is difficult to give a
clear picture of the practical complexity of results.   For that reason,
this section is devoted to discussing typical macrostructures with different types of welds in different materials.   In each case, the results
obtained by welding a material showing no phase change will be
compared with those obtained by welding a material showing a phase
change, usually a steel, illustrating as far as possible the difference
between unaffected parent metal, heat-affected parent metal and weld
deposit where applicable.   Description of the structure is confined
mainly to the figure captions to save space.

3.7.1   *Metal Structure in Liquid–Liquid Diffusion*

Plate 3.4 (a) illustrates a fusion weld single-V butt joint in copper,
made with a relatively low rate of heating and cooling (the oxy-
acetylene welding process); Plate 3.4 (b) shows a similar single-V butt
fusion weld in a mild steel.   For comparison Plate 3.5 shows fusion
welds made with high rates of heat input and fast cooling.   Plates 3.5 (a)
and in 3.5 (b) show similar welds to those in plate 3.4 but made by arc
welding whilst in 3.5 (c) is shown an electric resistance 'spot' weld in
aluminium (also a high rate of heat input and a high rate of cooling)
and in 3.5 (d) an electric resistance spot weld in a low-carbon steel is
shown.

In Plate 3.4 (a), the grain growth caused by the heat in the heat
affected zones can be seen, whereas, in Plate 3.4 (b), although no
marked grain growth is apparent, the structures of the heat affected
zones obviously differ from that of the unaffected parent metal and
from that of the weld deposit (which is not of identical composition
with the parent metal).   Plate 3.5 (a) shows less grain growth than in
Plate 3.4 (a).   The smaller extent of the heat affected zone of Plate
3.5 (b) is noticeable compared with Plate 3.4 (b) and traces of martensite can be seen.   In Plates 3.5 (c) and 3.5 (d), the typical cast structure
of each weld deposit is clearly visible, but there is very little heat
affected zone.

3.7.2   Metal Structure in Liquid–Solid Diffusion Welds

Plate 3.6 illustrates at (a) a silver brazed butt joint in copper sheet, at (b) a copper brazed joint in mild steel, whilst (c) shows a braze welded single-V butt joint with brass on copper sheet and (d) a braze welded T-joint in steel sheet, all made by the oxy-acetylene welding process. Compare these with illustrations in Plates 3.4 and 3.5 and note the effect on the heat affected zone structure, of the lower heat inputs required by the lower melting temperature weld deposits.

3.7.3   Metal Structure in Solid–Solid Diffusion Welds

Plates 2.6 and 3.7 show electric resistance-heated solid-phase butt welds in copper and mild steel respectively.   The amount and spread of deformation around the weld line is dependent on the heating intensity relative to the pressure and to the distance of the clamps from the weld face (the clamps are usually water-cooled and serve to conduct heat away from the weld) so the general contours may vary a good deal in practice.   Plate 2.6 may be compared with the macrostructures of Plates 3.4 (a) and 3.5 (a).

3.7.4   Metal Structure in Cold Pressure Welds

Cold pressure welding, that is solid-phase welding without externally applied heat, is not used for many applications, but it is interesting to note its main features.   The macrosection of a typical example of such a weld in aluminium is shown in Plate 3.8 (a).   The extensive deformation is obvious and the lack of intergranular growth across the interface can be seen in the microsection shown in Plate 3.8 (b).   The distortion of the interface indicates how mechanical 'keying' (see Section 2.6) can develop and the massive deformation shows how the intimate contact required to develop adhesion (also see Section 2.6) is attained.

3.7.5   Metal Structure in Soft Soldering

Soft soldering is a joining method relying almost entirely on adhesion for its strength.   In the macrostructure shown in Plate 3.9 (a), there seems to be a clear demarcation between the parent copper and the soft solder alloy (Lead–Tin–Antimony), but intermediate alloying is visible in the microstructure shown in Plate 3.9 (b).

3.8 EFFECTS OF POST-WELD HEAT TREATMENT OF WELDED JOINTS

None of the effects of heat treatment applied immediately after a welding operation differs from the effects of similar treatments applied to the unwelded material in similar initial states; but more than one influence may operate at once, since a variety of states may exist side by side in the vicinity of the weld.

Post-weld heat treatments, other than heat treatments not directly related to the use of welding (e.g., hardening and tempering a finished weldment in a heat treatable steel) are usually grouped under three headings (*a*) stress relief, (*b*) stabilisation of structure and (*c*) reformation of structure, and may be applied for one or more of the purposes indicated in these headings. The basic principles underlying the attainment of these purposes are summed up in Section 2.5.4, but their particular relevance to welding is considered briefly in the following sections.

3.8.1 *Post-Weld Stress Relieving*

The residual stresses left by welding (see Section 3.1.1 and Vol. 3, Chapter 3) can be quite severe and may give rise to serious problems of multi-axial stressing if they are left in until the weldment goes into service. The simple solution, if it is practicable, is to anneal the whole weldment, but many things may make this impracticable. For example, a large weldment requires a large furnace which may not be available; or the full annealing treatment may destroy a particular surface finish or desirable conditions in parts of the material away from the weld, or it may cause dangerous grain growth in or near the weld. Two alternative treatments are possible, either (*a*) to give the whole weldment a low temperature, or a sub-critical stress-relieving heat treatment (see Sections 2.1 and 2.5.4) or (*b*) to give a local low temperature stress-relieving heat treatment in the immediate vicinity of the weld. If a suitable furnace of temporary or permanent construction (it is often possible to make up a simple furnace on a temporary basis) is available, the first alternative is usually simpler and safer, because local heating has to be done very carefully to avoid uneven treatment, overheating and other undesirable effects, and, therefore, it requires more skilled metallurgical supervision.

For local heating purposes, the heat may be applied in one or more of several different ways, such as by portable hand torches (least reliable), automatic traversing torches, or fixed, specially located, burners (each using combustible fuel) or by low frequency electric induction heating or resistance heating from suitably located coils or resistors.

Some indication of the possible effects of low temperature stress-relieving on residual stress in a steel is given in Fig. 3.7. It must be noted that very long treatment times, even at the maximum treatment

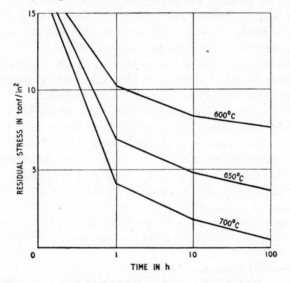

Fig. 3.7. *Typical effects of stress relieving a low-alloy steel at different temperatures*

temperature, are needed for anything like complete stress relief since the relief mechanism is mainly that of slow plastic strain within the microstructure.

Although a little out of context here, it is worth mentioning that local heating may be used to give a similar result to overload stress-relieving. In the latter case, a deliberately applied overload of the same type as the service load, is applied to cause plastic overstrain in the areas of critical residual stress, so that when the load is released the overstrained areas have either lost their initial residual stress or have it suitably reduced. It should be remarked, however, that to get such

102 WELDING FABRICATION

overload by heating requires even more careful control than the normal heat treatments.

3.8.2 Post-Weld Stabilisation of Structure

The rapid cooling usually encountered in welding often means that the weld structure is in a metastable state of supersaturation or of 'coring' (see Section 2.4.1). The former may tend to give natural ageing perhaps with resulting embrittlement and the latter to give uneven strength and properties, although this is not always important. Again, the simple solution is full annealing, but this may cause grain growth, coarsening of the structure and undesirable loss of mechanical properties in other parts of the weldment. Such undesirable effects may have to be risked if the 'as-welded' state is more dangerous; but the treatment must then be controlled very carefully.

Materials such as mild and low-alloy steels are readily treated by normalising (see Section 2.5.4) to give a relatively stress-free, comparatively stable structure, as shown in Plate 3.10. This shows (at (a) and (b) respectively) the as-welded and normalised microstructures of a typical weld section. If a proper annealing or normalising operation is not practicable, it may be necessary to use a prolonged sub-critical heat treatment to give as uniform a structure as possible, but, at the same time, taking care to avoid the development of other undesirable conditions such as grain growth or the precipitation of a brittle phase.

3.8.3 Post-Weld Reformation of Structure

With certain materials, a totally different approach may be needed. For example, to eliminate martensite in the vicinity of a weld in a high-manganese austenitic steel, it is desirable to heat the weldment to about 800°C, into the fully austenitic condition, and then quench in water to prevent martensite reforming (see Section 3.6).

Post-weld reformation of the structure of materials such as mild and low-alloy steels is also often desirable and is relatively easy to effect by normalising as shown in Plate 3.10. Other materials too will respond to a full heat treatment. For example, a weldment of a solution treatable material can be given a full solution treatment, followed by ageing, if required. An undesirable elevated temperature precipitate, such as chromium carbide in austenitic stainless steel, can sometimes be redissolved by heating the material into an appropriate temperature range (see Section 3.6) higher than that which caused the original trouble and then cooling it rapidly.

Many materials, including most pure metals, cannot be structurally reformed by heat treatment alone, and the only effect of annealing would be to cause more grain growth.   Such materials may require special fusion welding techniques if they are to be joined by such means.   An example of this is in the welding of copper (see Plate 3.4 (a) and 3.5 (a)) in which grain growth occurs readily.   If a coarse grain structure is intolerable, an enlarged weld can be made as shown in Fig. 3.8 (a).   After welding, the joint is cold forged down to standard thickness, as shown in Fig. 3.8 (b), then heat treated for an appropriate

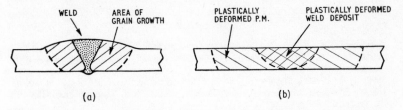

Fig. 3.8.   Grain growth refinement of a metal that does not refine with heat treatment alone, e.g., copper.   (a) As-welded.   (b) Hammered flat

time at a temperature just high enough to give recrystallisation of the cold worked material (see Section 2.1.4) without grain growth. Design of the joint and subsequent treatment have to be controlled rather carefully and, of course, the plastic flow during forging has to be accommodated.

3.9   WELDABILITY OF PLASTICS

Although the major concern of welding is with metals and alloys, it must not be forgotten that some non-metals, notably plastics, can be welded by processes based mainly on fusion techniques.

Thermosetting plastics cannot, at present, be fusion welded, but most thermoplastic plastics can be welded if heat is applied in a suitably controlled way and the joint position is accessible for welding manipulation.   Most thermoplastics can be welded without protection from the atmosphere provided that the temperature is kept below the charring temperature. Unfortunately, with manual welding processes this generally means using slow deposition rates. Distortion can be a problem but does not take such an aggravated form as in other materials where thermal conductivity is greater.

Welding heat may be derived either from a hot air jet (instead of a combustion flame) in a welding system similar to oxy-acetylene welding (q.v.) or from dielectric heating in systems analogous to spot or roller seam welding (q.v.) utilising radio-frequency types of energy generators. The former means can be used for welding quite thick sections up to perhaps $\frac{1}{2}$ in thickness, but the latter means is mainly applicable to relatively thin material, particularly, flexible plastic sheet such as that used for storage bags, protective clothing, etc.

BIBLIOGRAPHY

TWEEDDALE, J. G., *Metallurgical Principles for Engineers*, Iliffe Books Ltd., Chaps. 4, 9 (1962).
LANCASTER, J. F., *The Metallurgy of Welding, Brazing and Soldering*, Allen and Unwin (1965).
Studies of the Welding Metallurgy of Steels, British Welding Research Association (1960).
Welding Handbook, American Welding Society, 5th Ed., Pt. 1, Chap. 4; Pt. 3, Chap. 45 (1967).
Data on the Welding of Plastics, The Institute of Welding (1966).
TIPPER, C., *Brit. Welding J.*, **13**, 461 (1966).
CABELKA, J., and MILLION, C., *Brit. Welding J.*, **13**, 587 (1966).

Appendix

Table 1. MELTING TEMPERATURES, CHEMICAL SYMBOLS AND LATTICE STRUCTURES OF
THE BETTER KNOWN METALS

Metal	Atomic symbol	Stable lattice at room temperature	Melting temperature °C
Aluminium	Al	FCC	660·0
Antimony	Sb	Rhombohedral	630·5
Beryllium	Be	CPH	1,280 ± 40
Bismuth	Bi	Rhombohedral	271·3
Boron	B	Hexagonal	2,300 ± 300
Cadmium	Cd	CPH	320·9
Calcium	Ca	FCC★	850 ± 20
Chromium	Cr	BCC	1,800 ± 50
Cobalt	Co	CPH★	1,490 ± 20
Copper	Cu	FCC	1,083·0
Germanium	Ge	Diamond type	958·6
Gold	Au	FCC	1,063·0
Iron	Fe	BCC★	1,535·0
Lead	Pb	FCC	327·4
Lithium	Li	BCC	186·0
Magnesium	Mg	CPH	650·0
Manganese	Mn	Complex cubic★	1,242·2
Molybdenum	Mo	BCC	2,625 ± 50
Nickel	Ni	FCC	1,453·0
Niobium (Columbium)	Nb	BCC	2,000 ± 50
Platinum	Pt	FCC	1,773·5
Potassium	K	BCC	62·4
Silver	Ag	FCC	960·5
Sodium	Na	BCC	97·6
Tin	Sn	Diamond type	231·9
Titanium	Ti	CPH★	1,660·0
Tungsten	W	BCC	3,410 ± 20
Uranium	U	Orthorhombic★	1,690·0
Vanadium	V	BCC	1,735 ± 50
Zinc	Zn	CPH	419·4
Zirconium	Zr	CPH★	1,700·0

Metals marked ★ show lattice changes with temperature changes

Table 2. TYPICAL MECHANICAL PROPERTIES OF SOME LESS COMMON METALS
(Denny & Kendall G.E.C., U.S.A.)

Metal and test temperature	Tensile strength 1,000 p.s.i.	Proof stress, (0·2%) 1,000 p.s.i.	Elongation, %
ZIRCONIUM			
Pure, annealed, RT	34	10	47
Commercial, annealed, RT	50	35	35
Commercial, 40% CW, RT	80	70	13
Zircaloy 2			
Annealed, RT	68	45	22
Annealed, 260°C	32	18	36
Hot rolled, RT	82	66	42
40% CW, RT	108	99	36
HAFNIUM			
Pure, annealed, RT	59	22	35
Pure, annealed, 260°C	38	12	51
Pure, hot rolled, RT	79	63	15
VANADIUM			
Commercial, annealed	70	63	32
Commercial, hot rolled, RT	91	—	22
Commercial, hot rolled, 600°C	40	—	38
Commercial, hot rolled, 1,000°C	7	—	50
NIOBIUM			
Commercial, RT	39	24*	49
Commercial, 550°C	32	10·5*	24
TANTALUM			
Commercial			
Sheet, annealed, RT	50	—	40
Sheet, worked, RT	110	—	1
Wire, annealed, RT	100	—	11
Wire, worked, RT	180	—	1·5
CHROMIUM			
Pure, annealed, 330°C	41	27	14
Pure, annealed, 1,070°C	9	9	104
RHENIUM			
Pure, annealed, RT	164	135	28
Pure, worked, RT	322	311	2
Pure, worked, 1,070°C	124	—	1
Pure, worked, 1,480°C	40	—	1

* Proportional limit.   RT = room temperature.   CW = cold worked

Table 3. HARDNESS VALUES

(Approximate Hardness Equivalent Values given to the nearest 0·5 average)

| Vickers | Brinell | Rockwell (HR) | | | Shore Sclero- |
| | | A | B | C | scope |
HV	HB				
50	50		0		
60	55	21	17		
70	65	28	33		
80	75	34	41		
90	85	39	48		
100	95	43	54		
110	105	45	60		
120	115	47	66		
130	125	48·5	72		
140	135	50	77		
150	145	49·5	81		
160	155	53	84		
170	165	54·5	87		
180	175	56	89		
190	185	57·5	91·5		
200	195	59	94	12	31·5
210	205	59·5	95·5	14	33·5
220	215	60	97	16	34·5
230	225	60·5	98·5	18	36·5
240	235	61	100	20	38
250	245	62		22	39
260	255	63		24	40·5
270	265	63·5		25·5	41·5
280	275	64		27	43
290	285	65		28·5	44
300	295	66		30	45·5
310	303	66·5		31	46·5
320	310	67		32	48
330	318	67·5		33	49
340	325	68		34	50
350	335	68·5		35	51
360	345	69		36	52
370	353	69·5		37	53
380	360	70		38	54
390	370	70·5		39	55
400	380	71		40	55·5
410	388	71·5		41	57
420	395	72		42	57·5
430	405	72·5		43	58·5
440	415	73		44	59·5
450	423	73·5		44·5	60·5
460	430	73·5		45	61
470	438	74		46	62
480	445	74		47	63
490	453	74·5		47·5	63·5
500	460	75		48	64·5
510	468	75		48·5	65·5
520	475	75·5		49	66·5
530	483	75·5		49·5	67·5
540	490	76		50	68
550	498	76·5		50·5	69
560	505	76·5		51	69·5
570	513	77		51·5	70·5
580	520	77		52	71·5
590	528	77·5		53	72
600	535	77·5		54	73
610	542	78		54·5	73·5
620	548	78		55	74·5
630	555	78·5		55·5	75·5
640	561	78·5		56	76
650	567	79		56·5	76·5
660	573	79		57	77·5

(Continued on next page)

Table 3. *(Cont.)*

Hardness Systems					
Vickers	*Brinell*	*Rockwell (HR)*			*Shore Sclero-scope*
HV	HB	A	B	C	
670	577	79·5		57	78·5
680	585	79·5		57	79
690	590	80		57·5	80
700	595	80		58	80·5
710		80		58	81·5
720		80·5		58·5	82
730		80·5		59	83
740		81		60	83·5
750		81		61	84·5
760		81		61	85
770		81·5		61·5	85·5
780		81·5		61·5	86·5
790		82		62	87
800		82		62	88
810		82		62	88·5
820		82		62·5	89
830		82·5		62·5	89·5
840		82·5		63	90·5
850		82·5		63	91·5
860		82·5		63·5	92
870		82·5		63·5	92·5
880		83		64	93·5
890		83		64·5	94
900		83		64·5	94·5
910		83		65	95·5
920		83		65·5	96
930		83·5		65·5	96·5
940		83·5		66	97·5
950		83·5		66·5	98
960		83·5		66·5	98·5
970		83·5		67	99·5
980		84		67·5	100
990		84		67·5	
1000		84		68	

Index

111

Grain 7, 8, 41
Grain boundary 8
Grain-boundary fracture 85
Grain-boundary weakness 84
Grain growth 12, 14, 66, 78–79, 85, 94, 102, 103
Grain refining 58, 78
Grain size 8, 59, 89

Hardenability of low-alloy steels 90
Hardening, depth limitation 51
 precipitation 18
 quench. See Quench hardening
Hardness values 109, 110
Heat affected zone, effect of solid solubility 77
 failure in 73
 fusion welds 85
 precipitation during heating 78
 reaction with atmospheric gases 81
 structure in liquid–liquid diffusion 98
 structure in liquid–solid diffusion 99
Heat flow conditions of progressive fusion weld 74
Heat treatment 21
 post-weld 100–03
Heating, and weld structure 75–83
 gas absorption during 79
 grain growth during 78–79
 local 75, 101
 oxidation during 79
 precipitation during 78
 structural reactions during 82
 uneveness in 83
Hexagonal close-packed. See Close-packed hexagonal
Hot shortness 73
Hot tearing 71
Hot working 14
Hydrogen 47, 80, 82, 86, 92–94, 96
 atomic 92
 molecular 93
Hyper-eutectoid steel 50, 53
Hypo-eutectoid steel 50, 53

Impurities 85
Insolubility, complete 28
 solid 28
Interdiffusion 60, 65, 77
Interstitial solid solution 16
Inverse segregation 45
Iron 48
Iron carbide 47, 48
Iron–carbon diagram 48, 49
Isothermal transformation tests 53

Keying 60, 99

Lapped joint 65
Lattice stiffening by alloying 21
Lattice strengthening by supersaturation 20
Lattice structures 4–7, 107
Liquid solubility 28
Liquidus 25, 28

Macrostructure of welds 98
Magnesium 81
Manganese 95
Martensite 48, 51, 102
 decomposition control 52
 formation of 51, 90, 95, 96
 structure 51, 52
Martensitic steel 94
Mass effect 51
Mechanical breakdown of crystal structure 12
Mechanical properties 108
Melting temperatures 107
Metallic bond 5
Metallurgical basis to welding 2
Metallurgical bonding in welding 60–67
Metallurgical conditions in welding 73
Metallurgy 3, 4–67
Metals, crystalline nature of 4–15
 polycrystalline structure of 7
Metastability 16
Microcracking 86
Microstructure of welds 98
Mild steel 50, 91, 98, 99

Necking 21
Nickel 58, 95
Nickel equivalent 96
Niobium 95
Nitrogen 46, 80, 82, 87
Normalising 59, 102
Nucleus formation 8

Octahedral planes 7
Oxidation during heating 79
Oxy-acetylene welding 98, 99
Oxygen 46, 79, 80, 82, 87

Pearlite 50
Peritectic reaction 30
Peritectoid reaction 33
Phase 16, 17, 29